Praise for

# NEVER HOME ALONE

"Utterly fascinating... a spirited romp through the vast diversity that inhabits our daily lives and how we've changed our ecosystems, often for the worse.... This book will upend much of what we assume about the world around us."
—*Washington Post*

"In his fascinating new book... Mr. Dunn brings a scientist's sensibility to our domestic jungle by exploring the paradox of the modern home: In trying to make it 'clean,' we're forcing the species around us to evolve ever faster—often at our own expense. Mr. Dunn is a fine writer, wringing poetry out of the microbial explorations of Antonie van Leeuwenhoek, who spent half the seventeenth century documenting all the tiny living things around him.... Dunn tells us, 'all alone exploring a realm that was more diverse and elaborate than anyone but him seemed to understand.' Mr. Dunn also gracefully explains, without getting bogged down in details, the technology that has allowed scientists during the past decade or so to sequence the DNA of millions of previously unknown microbes, making his book an excellent layperson's guide to cutting-edge research."
—*Wall Street Journal*

"Chatty, informative.... It's hard not to be occasionally charmed by [Dunn's] prose, as when he catalogs the arthropods with whom we share our homes.... And it's hard not to share, at least a little, his awe at their diversity, even in a single household."
—*New York Times Book Review*

"[A] fascinating and illuminating book.... Dunn and his colleagues have used the concepts and techniques of community ecology to tease apart the functioning of a mostly ignored ecosystem: the human home. Their research enriches our understanding of ecosystem function, and—more grippingly—gives us insight into how our interactions with living things in the domestic habitat affect our health and well-being."
—*Nature*

"*Never Home Alone* is a thumping good book that raises alarm and offers reassurance in roughly equal measure. And it is funny.... What makes [it] so compelling is a sense of wonder and delight that encompasses all sorts of creatures and all sorts of science."
—*Los Angeles Review of Books*

"Intriguing.... Seen through Dunn's curious eyes, a house becomes not just a set of rooms, but a series of habitats to be explored. His writing and research lend a new appreciation of what many of us consider pests."

—*Science News*

"If you're an insectophobe looking for a thrill, you'll love Rob Dunn's *Never Home Alone*, which details the thousands of species of insects and microbes that live in and around your home." —*Bustle*

"If you could somehow infuse the curiosity of a six-year-old with PhD-level intelligence, imagine what wondrous things you could learn. Or why not make it easier on yourself, and just read *Never Home Alone*. Yes, that delightful, open-minded gee-whiz is exactly what makes this book so enjoyable. Surprisingly, it's doubly so for a germophobe, an arachnophobe, or anyone who can't stand the idea of intruders. Dunn has a way of brushing fears aside so he can tell you about something that's too cool to miss, or a fact that makes you say, 'Wow!'... Science-minded readers will love this book. It's filled with things you'll want to know for the health of it. Really, for anyone who's alive, *Never Home Alone* is a book to share with a few million of your newest best friends."

—Terri Schlichenmeyer, *Bookworm Sez* columnist

"If you enjoyed *I Contain Multitudes*, this book should be next on your reading list. Just like Ed Yong shows readers the fascinating microorganisms all around us, Dunn opens our eyes to the minute creatures that live within the confinement of our own homes.... A fascinating and entertaining read." —Read More Science

"Wonderful." —TreeHugger

"Crammed full of eensy-weensy tales of wonder from the insect world.... On virtually every page, readers learn about these marvels and their potential applications to the benefit of humans, all of it written with the bounce and insight of a true believer.... [A] hugely enjoyable book."

—Open Letters Review

"A robust, scientific defense for both microbial life and for larger creatures too often exterminated simply because they've invaded our space. While data from copious endnotes support staid scientific facts from strictly controlled lab tests, an engaging writing style enlivens narratives such as those about microbes in shower heads and beetles on windowsills, transforming

Dunn's latest work into a profound understanding of how all living things help in constructing and maintaining our planet's complex web of life."

—*Winnipeg Free Press*

"A lively compendium of hard science, anecdote, history, and personal memoir.... Something of a scientific raconteur, Dunn tells his story of the macro- and microbiome of our homes in a colloquial... style that makes the heavy science go down easy." —*Shelf Awareness*

"[An] intriguing and captivating scientific detective story.... Dunn eloquently observes that many species we find in our homes have value to us."

—*BookPage*

"Scintillating.... In a time of clear-eyed assessment of the environment, Dunn is a voice of reason who should be heartily welcomed."

—*Booklist*, starred review

"Of course we must chlorinate our water, wash our hands, get vaccinated, and so on, Dunn argues persuasively and entertainingly. But we also need to relax and cultivate biodiversity for the good of all life on Earth."

—*Kirkus Reviews*

"An entertaining tour of the biodiversity found in one of the fastest-growing biomes: indoors.... This book will be enjoyed by biologists but also general readers with an appreciation for nature." —*Library Journal*

"Delightfully entertaining and scientifically enlightening.... [Dunn] makes a compelling case for the value of biodiversity, while also conveying the excitement of scientific investigation, demonstrating that important discoveries can be made very close to home." —*Publishers Weekly*

"Easy to read and accessible.... Recommended." —CHOICE

"*Never Home Alone* is a superb guide to your own house—a place that is home to hundreds of thousands of species—a far richer habitat than even the largest backyard. This riveting and surprising book is one of those rare volumes that will make you gasp out loud on almost every page, and phone your friends to report stunning fact after stunning fact. Thank you, Rob Dunn: I love our house even more now I understand it shelters multitudes of fascinating (and mostly benevolent) living creatures."

—Sy Montgomery, author of *The Soul of an Octopus*

"If you're looking for a guide to the teeming, tiny, tenacious creatures that share our bodies, our homes—and may one day well inherit our planet—you could not do better than Rob Dunn. *Never Home Alone* is one fascinating introduction to the complex creatures surround us and, in ways that we are still coming to understand, make life on Earth riskier—but almost immeasurably richer."

—Deborah Blum, *New York Times*–bestselling author of *The Poisoner's Handbook*

"Rob Dunn is a brilliant explorer of the strange, mostly uncharted biology of our homes and bodies. This must-read book is full of astonishing stories, skillfully told."

—David George Haskell, author of *The Songs of Trees* and *The Forest Unseen*

"If you truly want to know yourself and be amazed, get to know your ecology. This charming book shows how important and fun it is to discover the astonishing world of marvelous and unseen creatures around us. You'll never take a shower again in the same way!"

—Daniel E. Lieberman, author of *The Story of the Human Body*

"In *Never Home Alone*, Rob Dunn reveals the unseen wilderness that surrounds us every day. This book will change the way you think about everything from dust to spiders to showerheads—a fascinating and highly recommended read!"

—Thor Hanson, author of *Buzz*, *Feathers*, and *The Triumph of Seeds*

"A Brooklyn couple visiting the West told me, 'We don't do nature.' Today, I'm sending them a copy of Rob Dunn's *Never Home Alone*, with this note: 'Nature made you, so roll in the dirt, open the windows, get a dog. Change your showerhead, but don't kill your spiders. And read this terrific book at once.'"

—Dan Flores, author of *Coyote America*

# NEVER HOME ALONE

ALSO BY ROB DUNN

*Never Out of Season*

*The Man Who Touched His Own Heart*

*The Wild Life of Our Bodies*

*Every Living Thing*

# NEVER HOME ALONE

*From Microbes to Millipedes, Camel Crickets, and Honeybees, the Natural History of Where We Live*

ROB DUNN

BASIC BOOKS
*New York*

Cover design by Rebecca Lown
Cover illustration © Olaf Hajek
Cover © 2018 Hachette Book Group, Inc.



Basic Books
Hachette Book Group
1290 Avenue of the Americas, New York, NY 10104

www.basicbooks.com

Printed in the United States of America

Originally published in hardcover and ebook by Basic Books in November 2018
First Trade Paperback Edition: October 2019

Published by Basic Books, an imprint of Perseus Books, LLC, a subsidiary of Hachette Book Group, Inc. The Basic Books name and logo is a trademark of the Hachette Book Group.

The Hachette Speakers Bureau provides a wide range of authors for speaking events. To find out more, go to www.hachettespeakersbureau.com or call (866) 376-6591.

The publisher is not responsible for websites (or their content) that are not owned by the publisher.

PRINT BOOK INTERIOR DESIGN BY JEFF WILLIAMS.

The Library of Congress has cataloged the hardcover edition as follows:

Names: Dunn, Rob R., author.
Title: Never home alone : from microbes to millipedes, camel crickets, and honeybees, the natural history of where we live / Rob Dunn.
Description: First edition. | New York, NY : Basic Books, 2018. | Includes bibliographical references and index.
Identifiers: LCCN 2018015934 (print) | LCCN 2018023036 (ebook) | ISBN 9781541645745 (ebook) | ISBN 9781541645769 (hardcover)
Subjects: LCSH: Biology--Popular works. | Natural history--Popular works.
Classification: LCC QH309 (ebook) | LCC QH309 .D866 2018 (print) | DDC

570—dc23

LC record available at https://lccn.loc.gov/2018015934

ISBNs: 978-1-5416-4576-9 (hardcover); 978-1-5416-4574-5 (ebook); 978-1-5416-1830-5 (paperback)

LSC-C

10 9 8 7 6 5 4 3 2 1

*To Monica, Olivia, and August, and all the species with which we have lived.*

# CONTENTS

*Prologue*

# *HOMO INDOORUS*

WHEN I WAS a child, I grew up outside. My sister and I built forts. We dug holes. We made trails and swung on vines. The indoors was a place reserved for sleeping, or for playing when it got so cold outside that our fingers seemed as though they might fall off (we lived in rural Michigan where that could happen well into the spring). But the outdoors—that's where we lived.

In the time since our childhood, the world has fundamentally changed. Kids today grow up primarily indoors, their lives punctuated by short bouts of movement from one building to another. This is not hyperbole. The average American child now spends 93 percent of his or her time in a building or a vehicle. Nor is it just American kids. The data are similar for children in Canada as well as in much of Europe and Asia.[1] I mention this not to bemoan the state of the world but instead to suggest that this transition reveals a radical new stage in the cultural evolution of our species. We have become, or are becoming, *Homo indoorus*, the indoor human. We now live in a world defined by the walls of our houses and our apartments, which are more connected to hallways and other homes than to the outdoors. In light of this shift, it seems as though we should make it a priority to know which species are living indoors with us and how they affect our well-being. But, in reality, we've only just scratched the surface.

We've known about the existence of other life in our homes since the earliest days of microbiology. At that time, it was studied by one man, Antony van Leeuwenhoek, who uncovered an astonishing number of life-forms in his home, on his body, and within the homes and bodies of his neighbors. He studied these species with a sense of obsessive joy and even awe. During the century after his death, no one really took up where he left off. Then, once it was finally discovered that some of the species in our homes could make us sick, the focus turned to those species, the pathogens. What followed was a huge shift in public perception, wherein we began to think of the species living alongside us, when we thought of them at all, as bad, as something we should kill. That shift saved lives, but it went too far, and as a result, no one ever really paused to study or appreciate the rest of the life in our homes. A few years ago, all of that changed.

Research groups, including my own, began to seriously reconsider the life in our homes. We began to study the life in homes in the way one might inventory a rain forest in Costa Rica or a grassland in South Africa. When we did, we were in for a surprise. We expected to find hundreds of species; instead, we discovered—depending on just how we do the math—upward of two hundred thousand species. Many of these species are microscopic, but others are larger and yet nonetheless overlooked. Breathe in. Inhale deeply. With each breath you bring oxygen deep into the alveoli of your lungs, along with hundreds or thousands of species. Sit down. Each place you sit you are surrounded by a floating, leaping, crawling circus of thousands of species. We are never home alone.

Just what kinds of species are living alongside us? There are, of course, the big species, the visible life. Around the world, tens, perhaps hundreds, of different kinds of vertebrates and even more kinds of plants can be found in homes. Far more diverse than the vertebrates and the plants, and still visible to the naked eye, are the arthropods, the insects and their kin. More varied than the arthropods, and often though not always smaller, are the species of the kingdom Fungi. Smaller than the fungi, and entirely invisible to the naked eye, are the bacteria. More species of bacteria have been found in homes than there are species of birds and mammals on Earth. Smaller still than the bacteria are the viruses, both those that infect plants and animals as well as the specialized viruses, the bacteriophages, that attack bacteria. We tally all of these different kinds of life independently. But the truth is they often arrive in our homes together. Our dogs, for instance, walk in our front doors carrying fleas in the guts of which live fungi and bacteria on which live bacteriophages. When the author of *Gulliver's Travels*, Jonathan Swift, noted that "each flea hath smaller fleas that on him prey," he didn't know the half of it.

YOU MIGHT, IN HEARING about all this life, be inspired to go home and scrub, and then scrub some more. But here is the other surprise. As my colleagues and I have looked at the life in homes, we have discovered that many of the species in the most diverse homes, the homes fullest of life, are beneficial to us, necessary even. Some of these species help our immune systems to function. Others help to control and compete with pathogens and pests. Many are potential sources of new enzymes or drugs. A few can help ferment new kinds of beers and breads. And thousands carry out ecological processes of value to humanity such as keeping our tap water free of pathogens. Most of the life in our homes is either benign or good.

Unfortunately, just as scientists have begun to discover the goodness, the necessity even, of many of the species in our homes, society at large has stepped up efforts to sterilize the indoors. Our increasing human efforts to kill the life in homes have unintended and yet very predictable consequences. The use of pesticides and antimicrobials, along with ongoing attempts to seal off homes from the rest of the world, tends to kill off and exclude beneficial species that are also susceptible to such assaults. In their place, we unknowingly aid resistant species such as German cockroaches and bed bugs and deadly MRSA bacteria (the methicillin-resistant species of *Staphylococcus aureus*). We not only favor the persistence of these resistant species—we speed their evolution. The evolution of species in our homes, alongside us, is arguably the fastest occurring anywhere on Earth. It may well be the fastest in the history of Earth. We are accelerating the rate of evolution in our homes at our own expense. Meanwhile, the vulnerable species that could compete with these newly evolved and ever more problematic strains are gone. Not to mention, the area affected by these changes is immense: the indoors is one of the fastest growing biomes on our planet, and it's now bigger than some outdoor biomes.

Perhaps it is easier to think about this shift in terms of a particular place. Let's consider New York and, within New York, Manhattan. In Figure P.1, you can see the amount of ground area in Manhattan. The larger circle is the ground area of floor, indoors. The smaller circle is the ground area of dirt, outdoors. The floor area indoors in Manhattan is now threefold greater than the dirt area outdoors. It is in this indoor world where any species able to survive finds huge quantities of food (our bodies, our food, our homes) and a favorable, invariant climate. Given such realities, the indoor world will never be sterile. It is sometimes said that nature abhors a vacuum. But that isn't quite right. It might be better to say that nature devours a vacuum. Any species that can colonize uncontested food and habitat will do so rapidly, like the tide coming in, sneaking beneath our doors and around each corner, crawling into our cabinets and beds. The best we can hope for is to populate the indoors with species that benefit us rather than do us harm. But if we are to do so, we first have to understand the species that have already made it indoors, those two hundred thousand or so species that we know so little about.

This book is the story of both the life that is likely living beside us in our homes and the ways in which that life is changing. The life in our homes speaks to our secrets, our choices, and our future.

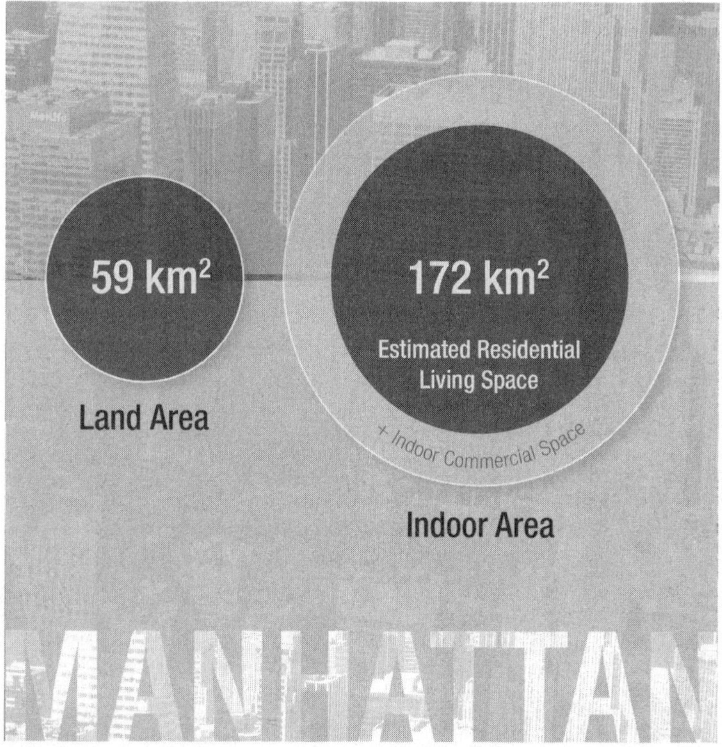


**Figure P.1** The indoor area in Manhattan is now nearly three times as large, in terms of its floor space, as the geographical area of the island itself. As urban populations continue to grow and densify, much of the world's population will soon be living in areas with more floor space than dirt. *(Figure adapted from NESCent Working Group on the Evolutionary Biology of the Built Environment et al., "Evolution of the Indoor Biome,"* Trends in Ecology and Evolution *30, no. 4 [2015]: 223–232.)*

It influences our health and well-being. It is full of mysteries and shimmers with grandeur and consequence. We do not know the stories of most of the species in our homes, but we know some of them, and what we do know will surprise you. When it comes to the species mating, eating, and thriving alongside us, nothing is quite what it seems.

# 1

# WONDER

> My work, which I've done for a long time, was not pursued in order to gain the praise I now enjoy, but chiefly from a craving after knowledge, which I notice resides in me more than in most other men. And therewithal, whenever I found out anything remarkable, I have thought it my duty to put down my discovery on paper, so that all ingenious people might be informed thereof.
>
> —**ANTONY VAN LEEUWENHOEK, in a letter dated June 12, 1716**

THERE IS NO singular origin story of the study of the wilderness of life in homes, but a day in Delft in 1676 comes close. Antony van Leeuwenhoek had walked the block and a half from his house to the market to buy black pepper. He strolled past the fish market, the butcher, and the town hall. He paid for the pepper, thanked the vendor, and then returned home. Once home, Leeuwenhoek did not sprinkle the pepper in his food. Instead, he carefully added a third of an ounce of the black stuff to a teacup filled with water. He then let the water and pepper steep. He was trying to soften the peppercorns so he could break them open and, in doing so, discover what it was inside them that led them to be spicy. Over the coming weeks, he checked on the peppercorns again and again. Then, after about three weeks, he made what was to prove a pivotal decision. He decided to draw a sample of the pepper water itself up into a thin tube of glass he had blown. The water seemed surprisingly cloudy. He examined it through a kind of microscope, a single lens affixed to a metal frame. This setup worked well for translucent things, such as pepper water, or for the thin sections of solid materials he would later teach himself to make.[1]

When Leeuwenhoek looked through his lens at the pepper water, he saw something unusual. Figuring out just what it was took some fidgeting and finessing. He either moved his candle this way and that, if working at night, or maybe he moved himself this way and that, if he was working using light from his window. He tried multiple samples. Then, on April 24, 1676, he finally had a clear view. What he saw was truly special: "an incredible number of very little animals of diverse kinds," as he put it. He had seen microscopic life before, but never anything quite so small. He would repeat this procedure in various permutations a week later, then again, then again with ground pepper, then with pepper in rainwater, then with other spices, each substance infused in his teacup. Each time he did this, he saw ever more life. These were the first sightings of bacteria by a human. And they were sightings being made in a home while studying materials that can be found in any kitchen, black pepper and water. Leeuwenhoek was at the edge of the wilderness, the miniature wilderness of his own home. He had seen a dimension of this living world that had never been seen before. The question remained whether anyone would believe what he had seen.

Leeuwenhoek probably started using microscopes to study the life around him, in his home and beyond, a decade prior in 1667. The moment when Leeuwenhoek saw bacteria in pepper water came only after hundreds, maybe thousands, of hours spent searching his home and daily life more generally. Chance does, indeed, favor the prepared mind, but it favors the obsessed mind even more. Obsession comes to scientists naturally enough. It emerges when one mixes focus and relentless curiosity. It can strike anyone.

Leeuwenhoek was not a scientist in the traditional sense. By trade, he worked with fabrics and sold cloth, buttons, and other related bits and pieces out of a shop in his home in Delft.[2] Leeuwenhoek likely began to use lenses of some kind to inspect the fine threads of particular fabrics.[3] But something then motivated him to explore other things in his home. It may have been a book published by Robert Hooke, *Micrographia*.[4] Leeuwenhoek spoke only Dutch, so he would have been unable to read Hooke's text, but the pictures of what Hooke had seen through his own microscope could have been inspiration enough.[5] From what we know of Leeuwenhoek's personality, it is easy to imagine him, after having seen the pictures, using the first Dutch-English dictionary (published in 1648) to puzzle through paragraph after paragraph of Hooke's words.

By the time Leeuwenhoek started to look through his microscope, other scientists had already used microscopes to see new details of home-dwelling creatures. Those scientists, including Hooke, found previously unsuspected patterns in life's interstices, patterns that suggested a world beyond that which was known. A flea's leg, a fly's eye, and the long-stalked spore cases (sporangia) of the fungus *Mucor* growing on a book cover in Hooke's home, all revealed minutiae not previously seen, or even imagined. We can examine the same species today, using the same magnification, but when we do our experience is very different from what it would have been in the 1600s. We already know that microscopic details exist even if we are surprised when we encounter them firsthand. For the scientists working in the early days of microscopy, the experience was more surprising, akin to discovering secret messages scrawled across each surface of the living world, messages no one else had ever seen.

As Leeuwenhoek peered through microscopes at the life in and around his home, he, too, saw new details. He saw the flea, for example, and drew many of the details that Hooke had drawn, but he also saw things Hooke had missed. He saw the flea's seminal vesicles, each no larger than a sand grain. He even saw the flea's sperm inside those vesicles, which he then compared to his own sperm.[6] As he continued to search, he began to notice entire life-forms that had never before been seen, life-forms entirely invisible without a microscope. These weren't overlooked details; Leeuwenhoek had found something more significant: he had discovered what we now call protists, a grab bag of single-celled life-forms united only by their size. They divided. They moved. And there were many kinds, some larger, some smaller, some hairy, some smooth, some with tails, some without, some attached to surfaces, and others unmoored.

Leeuwenhoek told people he knew in Delft about his discoveries. He had many friends, be they fishmongers, surgeons, anatomists, or nobles. One of those friends was Regnier de Graaf, who lived not far from Leeuwenhoek. De Graaf was a young man and yet already very accomplished. By the time he was thirty-two, he had discovered, for example, the function of the fallopian tubes. Leeuwenhoek's discoveries made such an impression on de Graaf that on April 28, 1673, he sent a letter to Henry Oldenburg, the secretary of the Royal Society in London, on Leeuwenhoek's behalf, despite the fact that he was mourning the death of a newborn child. In the letter, de Graaf noted that Leeuwenhoek had amazing microscopes and urged Oldenburg and the Royal Society to give Leeuwenhoek some specific assignments to pursue, subjects on which to focus with his microscope and skill. De Graaf also enclosed some of Leeuwenhoek's notes about his discoveries.

Upon receiving the letter, Oldenburg wrote back directly to Leeuwenhoek and asked him for figures to accompany his descriptions.[7] In August, Leeuwenhoek responded (by which time de Graaf had tragically died), adding more details about the things he had seen but that others (including Hooke) had missed: the physical appearance of mold, the stinger of a bee, the head of the bee, the eye of the bee, the body of a louse. Meanwhile, Leeuwenhoek's first letter, the letter that de Graaf had shared on his behalf, was published on May 19 in the *Philosophical Transactions of the Royal Society*, the second oldest scientific journal in the world, at that time still in just its eighth year. This was to be the first of many letters, letters akin to what one might now find in a blog post. The letters were not heavily edited; nor were they always structured. They were often digressive and repetitive. But these daily observations of the small things in his house and town were novel; they were observations of scenes no one had ever seen before. It was in one of these letters, sent on October 9, 1676, letter eighteen, that Leeuwenhoek recorded his observations about the pepper water.[8]

LEEUWENHOEK SAW PROTISTS in the pepper water. Protists include many kinds of single-celled organisms, each more closely related to animals, plants, or fungi than to bacteria. Leeuwenhoek described what appear to have been protist species of the bacteria-feeding genera *Bodo*, *Cyclidium*, and *Vorticella*. *Bodo* has a long whip-like tail (flagellum), *Cyclidium* is covered with wiggling hairs (cilia), and *Vorticella* attaches itself to surfaces by a stalk (and filters water for food). But then he also spotted something else. The smallest of the organisms in the pepper water were, he calculated, one one-hundredth the width of a grain of sand and one-millionth the volume. In retrospect, we know something so small could only be a bacterium. But in 1676, bacteria had never been seen by humans before; this was their grand reveal. Leeuwenhoek was thrilled, as he was quick to note to the Royal Society,

> this was among all the marvels that I have discovered in nature the most marvelous of all, and I must say that, for my part, no more pleasant sight has yet met my eye than this of so many thousands of living creatures in one small drop of water, all huddling and moving, but each creature having its own motion.[9]

The Royal Society had been pleased with Leeuwenhoek's first seventeen letters. However, with the letter on pepper water, he had finally gone too far, strayed from the path of truth toward that of pure imagination. Robert Hooke in particular balked. Hooke, thanks to the success of *Micrographia*, was the acknowledged king

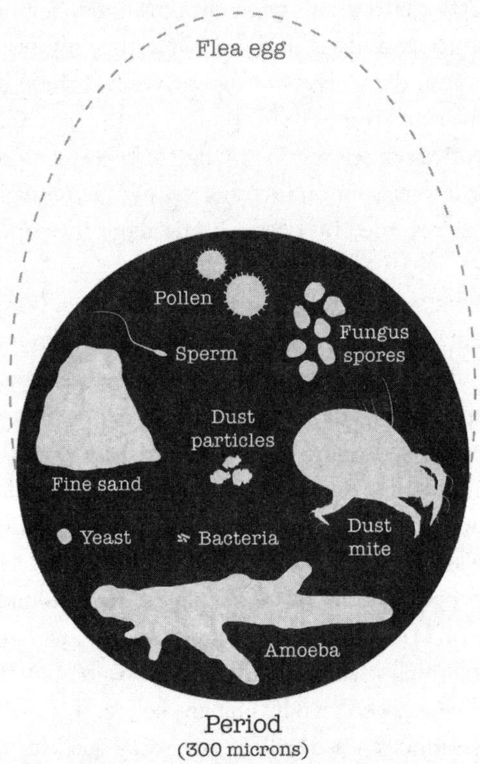


**Figure 1.1** Various life-forms and particles observed by Leeuwenhoek under his microscopes, drawn to scale relative to the period at the end of this sentence. *(Figure by Neil McCoy.)*

of the microscopic and had never seen anything so small alive. Hooke and another well-established member of the Royal Society, Nehemiah Grew, proceeded to try to repeat Leeuwenhoek's observations with an eye toward proving them false. It was part of what the society did, stage and repeat experiments. Usually they were done as simple demonstrations. In this case, however, the experiment was undertaken both as a demonstration and to determine whether or not the results Leeuwenhoek reported were true.

Nehemiah Grew was the first to try to repeat Leeuwenhoek's observations. He failed. Hooke took it upon himself to try. Hooke repeated each of the steps Leeuwenhoek took with the pepper, water, and microscope and he saw nothing. He grumbled. He scoffed. But he also tried again. He tried harder. He made better microscopes. On his third attempt, he and ultimately the other members of the Royal Society began, finally, to see some of what Leeuwenhoek had seen. In the meantime, Leeuwenhoek's pepper water letter, which had been translated into English by Oldenburg, was published by the Royal Society. With the publication of this letter, and the confirmation of Leeuwenhoek's observations by the Royal Society, the scientific study of bacteria—bacteriology—began. Notably, it began with the study of a bacterium found in a mix of ordinary kitchen pepper and water, a bacterium found indoors.

Three years later, Leeuwenhoek repeated the pepper experiment, but this time he kept the pepper water in a sealed tube. In the tube, the bacteria used up the oxygen that was present and yet something continued to grow, and bubble. Leeuwenhoek had once again discovered something new with the pepper water, this time the existence of anaerobic bacteria, bacteria able to grow and divide without oxygen. He once again made this new discovery while studying the life in his own home. The study of bacteria in general and the study of anaerobic bacteria in particular both began with the study of the life in a house.

We now know that bacteria are everywhere—in places with and without oxygen, in hot places and cold places, in every place—a layer, sometimes thin and sometimes thick, of life on each and every surface, inside each and every body, in the air, in the clouds, and at the bottom of the sea. Tens of thousands of bacterial species have been identified and millions (perhaps trillions) of other species are thought to exist. But in 1677, the bacteria Leeuwenhoek and a few members of the Royal Society had discovered were the only bacteria known in the world.

Leeuwenhoek's work is sometimes discussed, both historically and today, as though the man simply used a new tool to study the world around him and, in doing so, revealed new worlds. In this telling, the story is all about the microscope and its lens. The reality is more complex. Today, you can fasten a microscope of the same magnification as Leeuwenhoek used to your camera. (And you should.) If you do, you can use it to search around your house, but you will not see the world the way Leeuwenhoek did. Leeuwenhoek's discoveries did not result simply because he possessed a diversity of very good microscopes with well-made lenses. The discoveries depended on his patience, persistence, and technical abilities. It wasn't the microscopes that were magical but rather the combination of the microscopes and his careful hands and wonder-filled mind.

Leeuwenhoek was better at seeing this world, in all its grandeur, than anyone else. But doing so took work that others considered to be impossibly hard. So the members of the Royal Society, despite having seen the world Leeuwenhoek discovered, failed to continue to study it in any real earnestness. After verifying Leeuwenhoek's observations of microbes, Hooke continued to look at microscopic life through his own microscopes for about six months. But then he was done. Hooke and other scientists left the new world to Leeuwenhoek. Leeuwenhoek was to become an astronaut of the miniature, all alone exploring a realm that was more diverse and elaborate than anyone but him seemed to understand.

For the next five decades of his life, Leeuwenhoek systematically documented each and every thing around him; he documented all of Delft and beyond (often through samples brought to him by friends), but especially the living contents of his house. Anything he stumbled across was fair game. He studied the water in gutters, the water in rain, the water in snow. He detected microbes in his own mouth, and then in his neighbor's mouth. He observed living sperm (again and again) and showed how it varied among species. He showed that maggots arose from the eggs of flies rather than spontaneously from filth. He documented, for the first time, a kind of wasp that lays its eggs inside the bodies of aphids. He noticed, for the first time, that adult wasps survive the winter by slowing down and going quiescent. Over his years of dedicated study, he saw many kinds of protists for the first time, the first storage vacuoles,[10] the banded patterns in muscles. He discovered organisms living in the rind of cheese, in wheat flour, everywhere. He searched, he saw, he wondered, he discovered, again and again and again throughout the fifty years of his ninety-year life. He was like Galileo, dumbfounded and inspired. But whereas Galileo had to satisfy himself with looking out at the universe and the movements of stars and planets as tests of his predictions, Leeuwenhoek could touch the world he had found. He could discover the life in water and then drink the water, the life in vinegar and then use the vinegar, the species on his own body and then go about his life.

Because it is hard to match Leeuwenhoek's descriptions of the life around him to the modern names of species, we can't tally just how many life-forms he might have seen, but it was certainly in the thousands. It is tempting to draw a straight line from Leeuwenhoek to the modern study of the life in homes, but this would be wrong. Upon Leeuwenhoek's death, the study of the life in homes for its own sake was largely abandoned. Though Leeuwenhoek inspired the masses, he had no true colleagues in Delft after the death of de Graaf.[11] His daughter may have worked with him during his later years, but she did not follow up on his observations after he died. While she was alive she kept his specimens and microscopes, but they went unused. After she died, as Leeuwenhoek himself had specified in his will, they were auctioned off. Most of his microscopes disappeared. The gardens where he made observations were subsumed by the growing edges of Delft. His childhood home, where his inspiration must have first flourished, fell into disrepair and was torn down in the nineteenth century; in its place now stands a playground for a school. The house in which he made so many discoveries was torn down too.[12] A plaque was mounted to note the place where his house stood, but it was set in the wrong place. Another plaque was placed to remedy the error of the first; it, too, is not quite in the right location (one or two houses off depending on how one counts).

Eventually, other scientists would begin to study the life on human bodies and in homes anew. But by the time this happened, more than a hundred years had passed and it had been discovered that some microbial species could cause disease. These species were called pathogens. The idea that pathogens cause disease is the germ theory, credited to Louis Pasteur (though by the time Pasteur demonstrated that microscopic species could cause human disease it was already established that microscopic species could cause diseases in crop plants). With the advent of germ theory, pathogens became the focus of studies of microbial life indoors. Leeuwenhoek seems to have had an inkling that microscopic species could cause problems (he'd shown that some microbes could turn good wine into bad vinegar). He just imagined that most of the life he was seeing was harmless. In this, Leeuwenhoek was right. Of all the bacterial species in the world, for instance, fewer than fifty regularly cause disease. Just fifty. All the rest of the species are either benign or beneficial to humans, as are nearly all protists and even viruses (viruses wouldn't be discovered until 1898, though they too were discovered in Delft). Once pathogens were known to be part of the invisible world, war was declared upon all invisible life indoors. The closer that life was to us, the more exhaustive the war. The study of peppercorns, gutter water, and the whimsical, whirling creatures found everywhere in the average home was abandoned. Time would make this abandonment ever more complete.

By 1970, nearly the only studies being done in homes focused on pathogens and pests and how to control them. The microbiologists who studied homes studied how to kill pathogens. Nor was it just the microbiologists. The entomologists who studied homes studied how to kill insects. The plant biologists who studied homes studied how to get rid of pollen. The food scientists who studied pepper considered whether it might be a source of food-borne illness. We forgot about the potential of the life around us to inspire wonder and left no room for the realization that the species around us might not only plague us but also help us. We became focused on only part of the story. This was a big mistake, one we have only very recently begun to remedy. The first big steps back toward a more holistic view of the life around us were taken at hot springs—in Yellowstone National Park and in Iceland—places that seem to have nothing to do with homes at all.

# 2

# THE HOT SPRING IN THE BASEMENT

> Let both curiosity and horror—the latter of which terrorizes us but also holds us rapt, unable to look away—be a motivator for discovery. Embrace the weird, the tiny, the things we'd like to ignore.
>
> —BROOKE BOREL, *Infested: How the Bed Bug Infiltrated Our Bedrooms and Took Over the World*

IN THE SPRING of 2017, I was in Iceland filming a documentary about microbes.[1] As part of the filming, we stood, again and again, beside bubbling, hot, sulfurous geysers where I was meant to point at the geysers and talk on camera about the origins of life. At one point, I was even abandoned at such a geyser, only to wait for the truck to come back to get me.[2] Film crews can be unforgiving. While stranded, I had a moment to myself to contemplate the geysers. It was a cold day, and although they smelled sulfurous, I stayed near them. They kept me warm. The water in the geysers was boiling up from fissures in the Earth, warmed by the volcanism beneath the Earth's crust. In some places, it is easy to forget that the Earth is tectonic, just as one can become numb to the night sky. Not in Iceland. The western and eastern halves of the island are tearing apart and the consequences of this great rip of stone and dirt are hard to miss. Sometimes volcanoes erupt so violently that they darken the sky. And every single day geysers, like those I stood beside, bubble up out of the ground. As they do, they sustain life, life that has far more to do with what is happening in your home right now than you'd ever imagine.

That species survive and thrive in the warmth of geysers was not discovered until the 1960s. Thomas Brock, then at Indiana University, worked in Yellowstone and then also in Iceland, not far from where I stood. Brock was fascinated by the colorful patterns around the geysers. A smeared palette of yellow, red, and even pink gave way to greens and purples. Brock thought these patterns to be the work of single-celled organisms.[3] They were. The species present included bacteria but also archaea. The archaea are an entirely separate domain of life, as ancient and unique as the bacteria themselves.[4] What was more, Brock discovered that many of the species in the geysers were "chemotrophs," species able to turn the chemical energy of the geysers into biological energy; they made life from nonlife without the aid of the sun.[5] These microbes were of the sort likely to have existed long before photosynthesis ever evolved, their communities akin to some of the first communities. They evoked Earth's most ancient biochemistry. I could see them growing in a crusty mat around the geysers keeping me warm.

But these were not the only organisms in the geysers. Cyanobacteria were living in the hot water and photosynthesizing. In addition, Brock found bacteria that lived off of the organic matter swirling around in the bubbling water, be it the cells of other bacteria or a dead fly. Superficially, these scavengers were not very interesting. Unlike the chemotrophic bacteria Brock was studying, they couldn't turn chemical energy into life and instead had to find and consume living and dead bits of other species. However, after some study Brock decided that they belonged to a new species and even a whole new genus. He called the genus *Thermus* for obvious reasons and the species *aquaticus* to reflect its habitat.

For mammals or birds, finding a new species is a newsworthy event and finding a new genus, an even bigger deal.[6] But not for bacteria. It isn't hard to find new kinds of bacteria, and in terms of the features that microbiologists focus on first, this new species, *Thermus aquaticus*, didn't appear terribly interesting: it didn't form spores. Its cells were yellow rods. It was gram-negative. All true, all mundane. But there was something else.

Brock saw *Thermus aquaticus* in the lab only when he kept the growing medium (his cultures) at temperatures above 70 degrees Celsius (158 degrees Fahrenheit). It preferred even hotter temperatures and could still live at temperatures as high as 80 degrees Celsius (176 degrees Fahrenheit). The boiling point of water, for context, is 100 degrees Celsius, lower at higher elevations. Brock had grown what were among the most heat-tolerant bacteria on Earth.[7] As he would later note, finding this life-form wasn't hard. It was just that no one else had tried to grow microbes at temperatures so high. Laboratories had cultured samples from hot springs in 55 degrees Celsius culture conditions, too cold for *Thermus aquaticus* to grow well. Subsequent research has revealed an entire world of bacteria and archaea that can be grown only under very hot conditions. To such microbes, the temperatures at which we live out our ordinary lives are so cold as to be unlivable.

Why bring the story of *Thermus aquaticus* up in a book on houses? Because the temperatures and conditions found in geyseres and other hot springs, as unusual as they might seem, are very similar to those found around us in our daily lives. A student in Brock's lab thought it possible that *Thermus aquaticus* or other similar bacteria were even living, unnoticed, alongside us. To test the idea, the student and Brock probed the coffeemaker in Brock's lab, a machine that was plenty hot enough to favor *Thermus*. Given how much the machine helped fuel their work, it would have been an apt place to find the species. It wasn't there.

Brock found himself wondering about other places around him that contained hot liquids, such as the human body. Human bodies are not nearly as warm as hot springs, but Brock thought perhaps the bacteria might be present anyway, holding out for moments of fever. Who knew? It was easy enough to check. So Brock "produced" a sample of human spit (in an email, he declined to note whether it was his own, which in my experience studying the behavior of scientists means it was). He tried to grow *Thermus aquaticus* from the spit. Nope, no *Thermus aquaticus*. He checked human teeth and gums (much as Leeuwenhoek might have). None there either, nor any other heat-loving bacteria. Neither were there any in the lake from which he took a sample, nor in the nearby reservoir. He also checked the cactus in the greenhouse in his building, Jordan Hall. Nothing. Perhaps it really was a bacterial species found only in hot springs.

Just to be sure, Brock checked one more location: the hot water tap in his lab in Jordan Hall. Brock's lab was two hundred miles from the nearest hot spring. Yet, the lab's tap water contained what looked to be *Thermus aquaticus*. This was fantastic. Brock wondered whether the hot water heaters provided the habitat for the microbe—the water in the tap was warm, but not like in a hot spring. The hot water heater itself should be nearly perfect. Maybe the bacteria lived in the hot water heater and every so often, inadvertently, rode downstream to the tap.

Eventually another pair of researchers, Robert Ramaley and Jane Hixson, both of whom also worked at Indiana University, did additional sampling of thermophilic bacteria around Jordan Hall. When they did, they too found a kind of thermally tolerant bacteria. It was similar to the *Thermus aquaticus* noted by Brock. But it wasn't identical, so they called it *Thermus* X-1 for the time being.[8] Unlike *Thermus aquaticus*, it wasn't yellow. It was clear. Also, it grew faster than did *Thermus aquaticus*. Ramaley speculated that perhaps it was a new strain of *Thermus aquaticus*. Maybe the yellow pigment of *Thermus aquaticus* was an adaptation that protected it from the sun out in exposed hot springs. Perhaps, having colonized water sources in the building, this strain might have lost the ability to produce the expensive and unnecessary pigment. Brock, who had by then moved to the University of Wisconsin, decided it was time to study the *Thermus* in buildings in more detail.

Brock, along with his lab technician Kathryn Boylen, looked in hot water heaters both in homes and in laundromats near the University of Wisconsin. In laundromats, hot water heaters are often larger and used more consistently than those in homes, such that they might be even more likely to house thermophilic microorganisms. At each site, Brock and Boylen unfastened the drain on a hot water tank and examined what was inside. In hot water heaters, like in hot springs, temperatures can get very hot. In addition, all tap water contains organic material, perhaps enough of it to sustain *Thermus aquaticus*.

Over a century ago, the ecologist Joseph Grinnell applied the term *niche* to describe the set of conditions a species needs to survive. The word *niche* derives from the Middle French word *nicher*, "to nest." It was first used in reference to the shallow recesses in ancient Greek and Roman walls in which a statue or other object might be displayed.[9] The niches were just the right size for the statues, much as the temperature and food resources in your water heater seem to suit the needs of *Thermus aquaticus*. But just because a species can survive somewhere doesn't mean it arrives. Scientists now distinguish between the fundamental niche of a species (those conditions in which it could live) and the realized niche (those conditions in which it does live). The fundamental niche of *Thermus aquaticus* includes hot water heaters, but whether it was realized was another question altogether.

It was. Brock and Boylen found that species of the genus *Thermus* live, in addition to in geysers exposed to magma and in the tap water in Jordan Hall at Indiana University, in the hot water heaters of houses and laundromats around Madison, Wisconsin. What was more, the bacteria found in those hot water heaters were tolerant of temperatures as extreme as any at which life had been found anywhere. Brock went to the ends of the Earth to find species of the genus *Thermus*. He could have made the same discovery around the corner from his laboratory in the back room of the Suds and More.[10]

Since Brock's work, no other scientists have yet published papers about the *Thermus aquaticus* in hot water heaters. A new species of *Thermus* was, however, discovered in hot tap water in Iceland.[11] It turned out to be the same pigment-less species Brock and Boylen found in hot water heaters, a species now called *Thermus scotoductus* rather than *Thermus* X-1.[12] A graduate student at Pennsylvania State University, Regina Wilpiszeski, has spent the last few years sampling hot water heaters to see whether this is the main species present in hot water heaters. It appears to be: she has found *Thermus scotoductus* in hot water heaters across the United States. In thirty-five out of a hundred hot water heaters Wilpiszeski sampled, she found *Thermus scotoductus*. Wilpiszeski's work is not yet done, but it already raises new questions. Why is this species present in hot water heaters and how does it get there? And why have the many other heat-loving bacteria able to live in hot springs not yet colonized hot water heaters? Why don't very old hot water heaters take on the colorful microbial complexity of hot springs? So far, none of these questions have been answered.

I suspect different species of heat-loving bacteria live in hot water heaters in other regions. It is easy to imagine the species found in hot water heaters in faraway New Zealand or Madagascar might be totally unique. We don't know. Much in the way that few followed up on Leeuwenhoek's efforts, the same has been true for those of Brock.[13] Wilpiszeski stands alone. We don't know whether *Thermus scotoductus* has any consequences for us or our hot water heaters (be they positive or negative). Nor do we know whether the *Thermus scotoductus* bacteria in hot water heaters might have some uniquely useful attributes; the same species collected from other habitats seems to be able to make toxic forms of chromium nontoxic, among other tricks.[14] But the story of *Thermus* has been key in the history of the study of life in our houses. It was an indication—indeed, the clearest reminder since

Leeuwenhoek's time—that the ecosystems in our houses are more diverse than we had thought, populated with far more on hand than the pathogens that have received so much focus. Moreover, the *Thermus* in the water heater spoke to the possibility that the conditions of the modern home could have invited species indoors that never used to live around us, species that had moved in unnoticed. Ultimately, the presence of *Thermus* in hot water heaters helped, slowly, trigger a broader search for life in homes. It inspired people like me to consider the possibility that *Thermus* was not alone but instead part of some much bigger story. One can find, in houses, conditions as cold as the coldest colds, as hot as the hottest hots. One can find a microcosm of the world's conditions. It was entirely possible that these microbes had found and colonized our indoor extremes, but that no one had looked for them. The next revolution in the study of the home awaited new techniques, techniques that would allow microbes to be identified even if they could not be cultured in Petri dishes, techniques that would prove to be dependent on the unusual biology of *Thermus* itself.

We have known for a while now that most species of bacteria cannot be grown in the lab; they are still "unculturable." We don't know what food or conditions they need, so even if we sample them, we never see them. This means that for most of the history of microbiology these species were also unstudiable unless a clever and persistent biologist made an unculturable species culturable by figuring out its needs. Such was the case with species of the genus *Thermus;* they went unseen until Brock tried to grow them at high temperatures. But our ability to see the unculturable life around us has recently changed. It is now possible to study, and understand species we have no idea how to grow. This is thanks in no small part to Thomas Brock's discovery of *Thermus aquaticus* and its kin.[15]

The main tool we now use to initially find and identify unculturable species is really a series of laboratory steps. Those steps often are called a "pipeline," where *pipeline* just means the steps need to happen in order.[16] Into the beginning of the pipeline, one inserts a sample. Out the other side comes a list of the species, be they living, dormant, or even dead, present in the sample. This pipeline is an approach worth understanding in more detail because we have come to use it again and again in our research.

The pipeline begins with the samples. Once in the lab, samples are put into small tubes that contain a drop of liquid. The samples might be dust, or feces, or water—anything that contains or might contain cells and DNA. The liquid includes soap, enzymes, and tiny round glass beads, each the size of a grain of sand, which help to break open cells, like cracking eggs, to get out their DNA, the bacteria's genetic code. The tube is then sealed, heated, shaken, and centrifuged. The heavy beads and many cell bits and pieces sink to the bottom of the tube. The treasure, the long strands of less dense DNA, rises to the top to be skimmed off the way you might pull a dead fly from the surface of a swimming pool.[17] All of this is pretty straightforward and can be done in an introductory biology laboratory with sleepy students, some of whom have likely ignored most of the instructions.

To identify the different organisms on the basis of the resulting DNA (which has been "extracted" from its cells), we need to read the DNA, a process scientists call sequencing. This is the tricky part. As opposed to microscopes, which make the thing you are looking at appear to be bigger, sequencing techniques make the invisible information in DNA intelligible by first making it more plentiful. The trick was how to make it more plentiful so that the nucleotides of the DNA, its genetic letters, could be read. All DNA, except that in viruses, is double stranded. Two complementary strands are joined by a sort of molecular zipper. It was understood quite early on that if the two strands of DNA could be (gently) unzipped, each strand could be copied, and that this might be repeated until there was enough DNA to work with and decode. The two strands of DNA can be separated using heat. That much was easy. Copying the separate strands of DNA then just required use of an enzyme called polymerase, the same enzyme cells themselves, including human cells, use to copy DNA. You could separate the two strands of DNA, add some polymerase, a primer (the bit of DNA that told the polymerase which section of DNA, which gene, to copy), and some nucleotides, and you'd be on your way. The problem was that temperatures hot enough to pull the two strands of DNA apart were also hot enough to destroy the polymerase. One clumsy, expensive, labor-intensive way around this problem was to add fresh polymerase and primers after each round of heating. This approach worked, but was painfully slow, slow enough that in studying bacteria it was still easier for most microbiologists to just focus on the subset of species that could be cultured and ignore the unknown, unculturable bacteria for the time being.

A solution was forthcoming. The solution was *Thermus aquaticus*. The polymerase of *Thermus aquaticus* works at high temperatures. More than that, it works *best* at high temperatures. This polymerase was exactly what was needed. Several years after *Thermus aquaticus* was discovered by Brock, it was realized that the polymerase of *Thermus aquaticus* (nicknamed "Taq") could be added to DNA at high temperatures and the DNA would be copied rapidly. The copying of DNA using thermally tolerant polymerases, a process called the polymerase chain reaction (PCR), may seem abstract, a minor scientific footnote. Yet it is at the heart of virtually every genetics test being done in the world, whether it is to identify a child's paternity or the bacteria in a dust sample. The bacterial lineage discovered in hot springs and hot water heaters, a lineage that inspires our quest for unusual life in homes, also provides the enzymes necessary to carry out this quest across modern scientific research.[18]

Just which gene scientists, technicians, or clinicians copy during the polymerase chain reaction, and how they decode the resulting copies of DNA, depends upon the goal of the study and the technology being used. Studies that attempt to identify all of the bacteria in a particular sample tend to copy a single gene, the 16S rRNA gene, which is so central to the function of bacteria and archaea that it has changed little over the last four billion years.

For that reason, scientists can count on the gene being present in any species of bacteria or archaea studied. The gene differs enough among species to allow them to be distinguished, but not so much that it becomes unrecognizable. As for the technologies used for decoding the many copies of this gene, they vary greatly. Some rely on adding labeled nucleotides (those genetic letters) into the samples that have been or are going to be copied. The nucleotides are labeled with substances that can be read by a sequencing machine. The machine begins by reading each copy of the primer, that beginning stretch of nucleotides, and then it reads the letters that follow. It does this for all of what might be billions of individual copies of DNA in a sample, yielding enormous data files in which the code of each and every bit of copied DNA is listed. Those copies are then lumped into groups on the basis of their similarity to each other, and the codes of those groups of sequences can then be compared to the genetic sequences of known species in databases from other studies.[19] The mechanics of this process are ever changing, but one thing about them is not. Every year they are cheaper and easier. Handheld sequencing devices are now on the horizon. (Indeed, they already exist but are prone to errors in reading the DNA. With time they will improve.)

Today, then, thanks in no small part to *Thermus aquaticus*, it is now possible to take a sample and process it through the "sequencing pipeline" in such a way as to identify which species, living and dead, are present in the sample. This can be done without ever seeing or growing any of the species in the sample. Biologists can identify the life in soil, seawater, clouds, feces, and anywhere else. Biologists can identify culturable species but also the many, many species we do not yet know how to culture. Such a reality seemed impossible, inconceivable really, when I was a graduate student. Today, it is ordinary.[20] About ten years ago my colleagues and I decided to use such techniques to study the life in homes. At the time it had become possible, and affordable, to take a swab of dust from a door frame, a drop of water from the tap, or even a piece of clothing from the closet and to identify nearly all of the species present in that sample by decoding the DNA in the sample. Leeuwenhoek held his single lens up to the life around him. We would run the life around us through the sequencing pipeline. When we began, we really had no idea what we would find. The results would prove surprising. They were surprising both in terms of the many species we found to be present and in terms of those that were missing.

# 3

# SEEING IN THE DARK

> The only true voyage of discovery...would be not to visit strange lands but to possess other eyes, to behold the universe through the eyes of another, of a hundred others...
>
> —CHARLES DARWIN

My quest to understand the life in houses has its roots in the rain forest. When I was an undergraduate student, I spent part of my sophomore year at La Selva Biological Station in Costa Rica. I was working with Sam Messier, a graduate student from the University of Colorado, Boulder, who was studying the termite species *Nasutitermes corniger*. Worker termites eat dead wood and leaves of the forest, foods full of carbon but low in nitrogen. To compensate for the nitrogen missing from their diet, the termites host bacteria in their guts able to gather nitrogen out of the air. Colonies of these worker termites and their queens, king, and babies are defended by soldiers with long nose cannons that expel a kind of turpentine on their enemies, primarily ants and anteaters. The nose cannons of these soldiers are so long that the soldiers are unable to eat on their own and so they must rely on nutrients given to them by workers or gathered from the air by bacteria. Some *Nasutitermes corniger* colonies have many of these needy, dependent soldiers, whereas others have few. Sam wanted to know whether colonies produced more soldiers after having been repeatedly attacked by anteaters. There was an easy way to test Sam's hypothesis: simulate the effects of an anteater attack on some termite nests and not on others. This simulation was to be my job. I had a machete and would go termite nest to termite nest, day after day.

For the young boy still lurking in my twenty-year-old self, this job was great. I got to wander a trail hacking at things with a machete. For the young scientist in me, it was far better. While working, I talked to Sam about science until she tired. At lunch and dinner, I talked to other scientists until they tired. Then, when there was no one left to answer questions, I walked. At night, I walked paths with a headlamp, a flashlight, and a backup flashlight.[1] The night forest was full of the sounds of life and the smells of life, but the only things that could be seen were those revealed by the light. It was as if the light, in revealing species, also created them. I learned to tell the difference between the eye shines of snakes, frogs, and mammals. I learned to recognize the silhouettes of sleeping birds. I learned to look patiently at leaves and bark where giant spiders, katydids, and insects that mimicked bird feces lurked. Some nights, I convinced a German bat scientist to take me netting for bats. I hadn't been vaccinated for rabies. He didn't care. I was twenty, I didn't care. He taught me how to identify the bats. I learned the nectar feeders, the insect feeders, and the fruit feeders. I encountered the giant, bird-eating *Vampyrum spectrum*, so big it would rip a hole right through the net. My observations, however anecdotal, allowed me to begin to come up with my own hypotheses. I fell in love with the idea that most of what was understandable was not yet understood. I fell in love with discovery, with the way in which the unknown could be revealed with patience under nearly any log or leaf.

By the end of my stay in Costa Rica, I'd helped Sam show that the termite colonies were able to make more soldiers when bothered more often with a machete.[2] That was the end of the study, but not of the influence of the experience on me. I spent much of the next decade in Bolivia, Ecuador, Peru, Australia, Singapore, Thailand, Ghana, and elsewhere, moving in and out of tropical forests, threading through them as though I was trying to sew together some big picture. I'd return to the temperate zone, to Michigan or Connecticut or Tennessee, and then someone would offer me an opportunity—a free plane ticket, a mission, and all the beans and rice I could eat—and I'd suddenly find myself once more in the jungle. In time, I found the same kinds of discoveries and joys I associated with the rain forest in other realms, be they deserts or temperate forests. I even began to find them in backyards. This shift to backyards started when a new student, Benoit Guenard, joined my lab. Benoit is fascinated by ants. When Benoit arrived in Raleigh, he searched the forests for ants, ceaselessly. He turned up a species neither he nor I could identify. It was an introduced species, the Asian needle ant, *Brachyponera chinensis*.[3] The Asian needle ant had become common in Raleigh without anyone really noticing. In studying this ant, Benoit realized it exhibited behaviors never before seen in insects. For example, when one forager finds food, rather than lay a pheromone trail for others to follow, she returns to the nest, grabs another forager, carries her to the food, and throws her down upon it: "Here, food!"[4] Benoit went to Japan to study the Asian needle ant in its native range. Once there, he found a totally new species of ant, related to the Asian needle ant, common but unnoticed across much of southern Japan, including in and around cities.[5] These discoveries were just the beginning.

At about this time, back in Raleigh, a high school student, Katherine Driscoll, came to the lab. Katherine wanted to study tigers. I didn't study tigers, and so Benoit and I told Katherine to go search for and study *Discothyrea testacea*, "the tiger ant." What we didn't tell Katherine was that we had just made up the name "tiger ant" and that no one had ever found a living colony of these ants. Katherine went off looking. I imagined she would get distracted as she searched and find something else on which to focus. Instead, Katherine found the "tiger ant"; not only that, she found it in the soil behind the building that houses my lab and office. She was, at the age of eighteen, the first person to ever see a *Discothyrea testacea*, "tiger ant" queen alive.[6] Soon, we started to engage even younger students in helping us to sample ants in backyards, but no longer just in Raleigh.[7] We made kits that allowed kids across the United Stated to sample ants in their backyards. When we did, the rate of discovery accelerated even more. An eight-year-old discovered that the Asian needle ant was found in Wisconsin. Another eight-year-old found it in Washington State. No one knew it had spread anywhere beyond the southeastern United States.

This work engaging kids in the study of backyard ants precipitated a change in the lab. We began to involve the public more often in helping us to make discoveries. At first it was tens of people, then hundreds, and soon thousands of people looking where they lived for discoveries. By following these discoveries—discoveries we were making with the public—we eventually began to study the life indoors. It was thrilling to work with people to find new species and behaviors in backyards in no small part because those discoveries were immediate to people's daily lives. We were reminding people of the mystery that remained in the world around them. We were, I hoped, offering a little of the thrill that I experienced when I was twenty in Costa Rica but that I might have also experienced where I grew up in Michigan if I'd known there were still discoveries left to be made. It would be even more exciting, we thought, for the people we worked with to find new species, behaviors, and other discoveries where they actually spend most of their time: in the wilderness indoors.

Most studies of indoor life focused on pests and pathogens, so it was easy to imagine that other species might have been overlooked. At the time, scientists had done individual studies here and there on interesting, nonpathogenic, and nonpest species in homes (such as the work on *Thermus scotoductus* bacteria in hot water heaters), but these were one-offs, small studies rather than larger dedicated works. There was no field station directed to the study of, well, the inside of the field station. I assembled a team to study houses, a team that has continued to grow ever since, a team that includes scientists around the world and the public—adults, families, kids. Together we would seek Leeuwenhoek's exhilaration, the exhilarating madness of the possible. We were nearly ready. But there was still one trick, figuring out where to start and how to see. We decided to start with bacteria. I'd been interested in the bacteria in nests since my work with Sam Messier on the *Nasutitermes* termites—and what is a house if not a big nest? It seemed likely that it was among bacteria and other microbes, species invisible to the naked eye, that the biggest discoveries would be made. But to study these species we were going to need more than a microscope with a single lens. Times had changed. This is where Noah Fierer, a microbiologist at the University of Colorado, Boulder (in the same department where Sam Messier was a graduate student), came in. Noah provided the tool through which we would see the life inside homes. He could identify the species present in dust on the basis of their DNA; he could sequence the life in dust and in doing so reveal the invisible life we walk through and breathe.[8]

By training and inclination, Noah is a soil microbiologist. He is fascinated by soil; in it, he finds the same wonder I find in jungles, a way to lose himself in discovery. Fortunately, he can also be intrigued (or maybe *distracted* is the better word) by life elsewhere, so long as it isn't bigger than a fungal spore. Start talking about an ant or a lizard and Noah's eyes glaze over. Regardless of the habitat in which he is studying small life, Noah has a genius, like Leeuwenhoek's, for using a common tool in new ways. Leeuwenhoek is often said to have invented the microscope, which isn't true. Nor is it necessarily even true that Leeuwenhoek had particularly special microscopes. Instead, what was special about Leeuwenhoek's microscopes was Leeuwenhoek. In the same way, what is special about Noah's investigations is not that he has great devices for decoding the identity of microbes in samples (though he does); it is the way he uses these devices and techniques to see what others have missed. Noah would identify the species present in samples from homes by sequencing the DNA present in those samples. Noah and members of his lab would extract the DNA from each sample, make more copies of that DNA using the enzymes from *Thermus aquaticus* (or, by then, some other thermophilic microbe), and then decode the genetic sequence of particular genes common to all of the species in the sample. In doing so, he could reveal not only those species that scientists know how to culture but also those that no one can. Together with the public and Noah, we would be able to detect everything, living or dead, dormant or dividing, in homes.

Our plan was to enlist the public to help us sample the dust from ten habitats in each of forty houses using cotton swabs. The houses would all be in Raleigh, North Carolina, the city in which I lived and live. We needed to start somewhere and we knew so little about the indoor environment that Raleigh was as good a place as anywhere. We chose to sample refrigerators—not the food in them but instead the growth alongside the food. We sampled the dust on door frames, both inside homes and outside. We sampled pillowcases on beds, also toilets and door knobs, and kitchen counters too. Rather, we had the participants sample all of these places.

We sent each participant[9] the cotton swabs that they were to use to sample the habitats in their homes. The dust on the used swabs would contain what Hannah Holmes has called "fragments of a disintegrated world": bits of paint, clothing, snail shell, couch fibers, dog fur, shrimp shell, marijuana residue, and skin. The dust would also contain bacteria, living and dead.[10] The participants would then seal the swabs in airtight tubes and send them to Noah's lab, where nearly every bacterial species in each and every dust sample would be identified. Noah's lab would be the light through which we saw the hidden life in dust.

I'M NOT SURE what Noah expected from this survey of houses, but I can tell you what was known in the scientific literature when we began, what had been learned since the work of Leeuwenhoek in

**Figure 3.1** Jessica Henley handling DNA samples soon to be spun down in a centrifuge, one of the steps required to prepare DNA that has been isolated from environmental samples for sequencing. *(Photograph by Lauren M. Nichols.)*

the 1600s. Beginning in the 1940s, studies had shown that bodily bacteria can be found around houses. Bodily bacteria thrive in the places humans spend more time, especially those places they touch with their naked skin, be they toilet seats, pillowcases, or remote controls. These studies focused on discovering problem species, the fecal bacteria in the cauliflower and the skin pathogen on the pillowcase, and their eradication. Anything that wasn't worrisome wasn't of much interest. More recent studies from the 1970s revealed other kinds of species in homes: *Thermus* in water heaters and unusual bacteria lurking in drains, for example. These newer studies hinted at the possibility that we might find many new life-forms as we explored homes. We did.

Across the forty houses, we found nearly eight thousand kinds of bacteria, roughly as many bacterial species as there are species of birds and mammals in all of the Americas. The species we encountered were not just well-known species from human bodies but also many other life-forms, some of them very unusual. We turned over the metaphorical leaves of forty houses and beneath them found a wilderness. Many of the species didn't match up with anything yet known to science. They were new species, or even new genera. I was ecstatic, back in the jungle again, albeit the jungle of everyday life.

We decided to engage more participants to sample more houses. It took a while, but we were able to convince the Sloan Foundation, which had by then begun an ambitious effort to fund studies of the life in homes, to pay for a broader study. We also persuaded an additional one thousand people across the United States to swab four sites in their houses.[11]

In samples from those one thousand houses, we once more identified the bacteria. One might expect that we would have seen, in this second set of houses, species similar to those we saw in Raleigh. We did, to some extent. Many of the species found in Raleigh could also be found in Florida houses and even Alaska houses. But we also found species that were not seen in Raleigh, new species in each house and in each region. We saw, in total, some eighty thousand kinds of bacteria and archaea, ten times more than in the first sample of Raleigh.

The eighty thousand species we found included species from nearly all the most ancient branches of life. Species of bacteria and archaea are grouped into genera which are grouped into families which are grouped into orders which are grouped into classes which, in turn, are grouped into phyla. Some phyla, while ancient, are very rarely encountered. Yet, in homes, we found nearly all of the bacterial and archaeal phyla so far known on Earth. We found phyla that, a decade earlier, were not even known to exist, and we were finding them on pillows or in refrigerators. Here, then, was a humble moment amid the grandeur of life on Earth and in life's history. To really make sense of the life in our homes, we would need to study, in detail, the natural history of tens of thousands of species. (We aren't there yet; we won't be for decades.) But even before we attempted that, we began to see broad patterns, ways of grouping this mass of life to make it a little more intelligible.

Some of the bacterial species we found in homes were those that had already received some attention: bodily bacteria. But most of these species were not pathogens but instead detritivores, living off the awkward reality that our bodies are slowly falling apart even while we are alive. We leave a cloud of life everywhere we go. As we wander through our homes, our skin flakes off in a process called desquamation. We all fall apart at a rate of about fifty million flakes a day. Each flake floating through the air has thousands of bacteria living and feeding on it. Riding their skin flake parachutes, these bacteria fall from us like a steady snow. We also leave bacteria on the bits of bodily fluids—saliva and more—and feces deposited here and there. As a result, the places where we spend time in our homes bear the marks of our presence. Every place we put our bodies in every house we have ever studied offers microbial evidence of lives lived.[12]

That we leave bacteria in our wake is not surprising. It is inescapable and largely harmless, or at least harmless in settings in which modern waste treatment facilities and a supply of "clean" drinking water (we will return to just what that means later on) are available. The vast majority of species that you or someone else leaves on a chair when you sit down are beneficial or benign species that, for a brief moment of time, eat whatever bit of you has fallen off, before dying. They are gut bacteria that help you digest your food and that produce necessary vitamins. They are skin bacteria that grow all over your body and help you fight off pathogens. They are armpit bacteria that help your body fight off pathogens when they arrive on your skin. Hundreds of studies have now considered this trail of microbes we leave wherever we go. You see these studies in the news. Human bacteria are found on cell phones, on subway poles, and on door handles. They are found everywhere we go in proportion to population density. They always will be and that is fine.

In addition to the species associated with the falling apart of our bodies, we also saw species associated with the decay of our food—rot. These species were, unsurprisingly, most abundant in the refrigerator and on cutting boards, but they were elsewhere too. One of the samples taken from a television was composed almost entirely of food-associated bacteria. Sometimes we are left to guess at what a sample like that means. Science is full of enigmas.[13] Regardless, if the species that rot our food and live off of the slow decay of our bodies were the only species we found in houses, it would have been scientifically unremarkable, akin to going to Costa Rica and "discovering" that the rain forest contains trees. But the microbes of bodies and rotten food were not the whole story, not hardly.

As we looked in more detail, we found other kinds of microbes, bacteria and archaea like those for which Brock might have searched, extremophiles, species that "love" and thrive in extremes. For an organism the size of an archaeon or a bacterium, your home contains incredible extremes. Most of these extremes are new habitats we have unintentionally invented. Homes contain refrigerators and freezers that can get as cold as the coldest tundra. They contain ovens hotter than the hottest desert and, of course, hot water heaters as hot as hot springs. But homes can also include very acidic conditions, such as in some foods (like sourdough bread starters), and very alkaline (basic) conditions, such as in toothpaste, bleach, and cleaning products. In these extremes of homes, we found species once thought to live only in the deep sea, on glaciers, or in remote salt deserts.

The soap dispensers in dishwashers appear to be a unique ecosystem filled with microbes able to survive hot conditions, dry conditions, and wet conditions.[14] Stoves contain bacteria able to live in extreme heat. Recently, one species of archaea has even been found to survive in autoclaves, the superhot devices used to sterilize equipment in laboratories and hospitals.[15] Long ago, Leeuwenhoek showed that pepper can contain unusual life-forms. We found salt does too. Freshly purchased salt contains bacteria typically found only in salt flats out in deserts and areas that were formerly oceans. Sink drains contain a mix of species seen nowhere else that includes both bacteria and tiny drain flies, whose larvae feed on the drain bacteria. (You see drain flies often, probably without realizing it. Their wings make a heart shape and each wing is patterned with what looks like lace.) The pipes of showerheads—pipes that get dry, then very wet, then dry again—are covered in films of unusual microbes seen typically in swamps. Such new ecosystems are often physically small. In addition, the niches of their species are often narrow. The species often require very particular conditions. As a result, they are easy to miss, just as species with narrow niches can be easy to miss outdoors as well. That "tiger ant" that Katherine found, for example, is hard to find because it lives only in the egg cases of spiders, spiders that hide their egg cases underground.

Nor was the life in extreme habitats the last big discovery we found in homes. There was something else: a set of species in some but not all homes, species that, though not always common, accounted for much of the total biological diversity we found. These were species associated with wild forests and grasslands, typically found in soil, on the roots of plants, on leaves, and even in the guts of insects. This wild biodiversity was most common on outer door sills, then on inner door sills, and then it turned up here and there in other habitats in some (but not all) houses. These species may be living in the air on the bits of soil and other substances on which they have come in. They may be quiescent, waiting for just the right food, or they may be dead. Just which of these outdoor species drift indoors seems to depend on what is outdoors. The wilder the life outside a house, the wilder the life drifting through the air and settling on the doors is.[16] It would be easy to think of these drifting wild species, the flotsam and jetsam of homes, as irrelevant trespassers. Easy but very wrong.

LET'S PAUSE HERE before I tell you the stories of the individual species you are breathing in right now and the stories of what happens in houses with lots of outdoor bacteria and the stories of the rest of life (the arthropods, the fungi, and more). Let me first put what we are finding in homes in a little bit more context. To really make sense of what is living with you in your home, you need to consider it relative to a longer history, the history of homes.

For most of human prehistory, we slept in nests built of sticks and leaves. We can infer this on the basis of the lives of modern apes. We share a common ancestor with these apes. Those characteristics that are different from one ape to the next say relatively little about our common ancestor, but those traits that are shared speak to habits our ancestors likely shared as well. All living apes build nests out of loosely interlaced sticks and leaves. Chimpanzees build them, as do bonobos, as do gorillas, as do orangutans.[17] Apes tend to use their nests for a single night and then abandon them. They use nests more like beds than like homes, beds constructed in ephemeral settlements sometimes described, quaintly, as "dormitories."

Recently, Megan Thoemmes, a graduate student in my lab at North Carolina State University, studied the bacteria and insects in chimpanzee nests. One might predict that these nests would be full of species associated with the bodies of chimpanzees, be they chimpanzee bodily bacteria or maybe even larger species that sneak in to take advantage of the chimpanzees. (Sloths, after all, have an entire ecosystem of arthropods and algae that live in their fur,[18] why not chimps?) Fur mites. Dust mites. Maybe hide beetles. Spider beetles too. This is what we find in human beds.[19] We humans are exposed, when we sleep, to the ecosystem of our decay. Instead, Megan found the nests of chimpanzees to be occupied, nearly exclusively, by environmental bacteria, bacteria from the soil and leaves.[20] Just which bacteria depended on whether Megan's samples were from the dry season or the wet season. It is likely that these were the same sorts of species that would have been found in the nests of our ancestors until they first began to build homes. The bacteria to which our ancestors were exposed for millions and millions of years would have been environmental, with the precise mix different from season to season and place to place.

When our ancestors needed something more permanent than nests, they may have initially moved into caves. Eventually, though, they began to build houses. The oldest evidence of a structure built by our ancestors is from a campsite near a beach at Terra Amata (near modern Nice).[21] There, an archaeologist found evidence of at least twenty homes along an ancient shore. The most intact of these homes showed a ring of stones surrounding a floor of ash. In the floor, the location of posts used to support the roof were still visible. Around the stones, in a second ring, were the marks of stakes, each of which apparently ran from the ground and bent in to form the room. These houses were built by an ancient hominid (probably *Homo heidelbergensis*) more than 300,000 years ago.[22] We know little about how common such houses were, how varied they were, or when they first appeared. The archaeological record offers us little, just clues here and there. For instance, shelters attributed to hominids (in this case, modern humans) at a 140,000-year-old site in South Africa. Beds at a 70,000-year-old site in South Africa.[23] Whatever was going on, at least some of our ancestors were sleeping indoors, separated a little from the world outside.

By twenty thousand years ago, house sites start to turn up around the world. In nearly every case, the houses appear to have been round and domed. They were simple, like the chamber that a termite king and queen, out on their own, might build for themselves. In some places, they were made of sticks; in others, mud. Others, still, in the far north, were made of mammoth bones. Some of these homes were more ephemeral than others, used for just a few days or weeks, but I suspect that already, in these homes, we had begun to alter the species present around us. The best evidence of this change comes from studies of modern peoples living in houses similar to those our ancestors used. In Brazil, for instance, traditional, open-walled, palm-roofed Amazonian houses built by the indigenous Achuar are dominated by environmental bacteria.[24] Similarly, Megan Thoemmes found that even though the houses of the Himba in northern Namibia were but a single round dome, the places where people slept contained different microbes than those where they cooked. Even simple houses tend to allow the buildup of body microbes. Yet, whereas the Himba and Achuar homes include body microbes, they are also, like chimpanzee nests, as diverse with environmental bacteria as is the air around the houses. Indoor microbes build up in contemporary Himba homes, and contemporary Achuar homes, but environmental microbes are still present as well. The modern homes of the Himba and Achuar are imperfect proxies for our historic dwellings. Yet it seems safe to suggest that exposures of our ancestors in homes like those found in Terra Amata, France, would have been similar to the Achuar and Himba homes in terms of the preponderance of environmental microbes.

Where once there were only round houses, humans began to build square houses roughly twelve thousand years ago. Though square houses had less usable area inside than did round houses, they were easier to make modular. Large numbers of houses could be stacked side by side or even on top of each other. The shift from round houses to square houses happened in nearly every place humans began to farm and live in greater densities. With this change, houses came to be slightly more isolated from the outside world. There was, to a greater extent, an inside and an outside. The old-style houses didn't go away though. Both round houses and square houses existed side by side.

Fast-forward twelve thousand years. Today, the vast majority of humans live in cities, a trend that is only accelerating, and, in cities, ever more people live in apartments. The distance an outdoor bacterium needs to travel to get inside an apartment can be lengthy. If the windows of the apartment stay closed, a bacterium must make its way up the stairs, down the hall, past some doors, and then quickly inside. We imagine we can create a world that is sterile. But in apartments with closed windows, where the route up from the park is far, what we instead create is a world filled primarily with microbes associated with our falling apart, our food's falling apart, and even the building's falling apart. Once, we lived in nests in which the microbes around us were all just environmental microbes and our imprint on the places we sat or slept was so modest as to be almost undetectable. Now, in some apartments, the imprint of the environment, of nature, is all but undetectable. But here is the key: our study's results show that life inside apartments varies as it does among houses. Some dwellings are really cut off from the environment; others, like those of the modern Himba or Achuar, less so. We have choices, choices as to how much of life's richness we allow in.

IN MY EXPERIENCE, upon learning they live among thousands of kinds of bacteria in their homes—be they detrital species, extreme species, or species of wild forests and soils—people have one of three responses. Microbiologists, whom I hang out with fairly often, are a little more impressed than they were initially, but still not overwhelmed. "Eighty thousand? I would have thought more. Did you sample in winter too? Did you swab a dog?" Microbiologists are immersed every day in the grandeur and vulgarity of the unknown; they get numb to it. Let's ignore the microbiologists for now.

Some people feel awe. I feel awe. Awe is what I hope to also inspire in others. It is awesome to walk about in a diversity we have yet to even begin to contemplate. It took four billion years for the microbial diversity we encounter in our homes to evolve. Each home is filled with unnamed species about which we know nothing; some we might have been living alongside for millions of years, others have more recently colonized the nooks and crannies of our modern lives. There are discoveries yet to be made all around you even without leaving home. New species. New phenomena. New everything.

But many people feel disgust. How do I know this? Because when we make discoveries in homes, we report those discoveries back to their denizens. When we do, people email us questions. I enjoy the questions. Sometimes they are like the questions I asked field biologists when I was at La Selva Biological Station in Costa Rica: What do we know about this species? What does it do? Often, the answer I can provide is similar to what the tropical biologists could provide me: "We don't know. You should study it." Or, "We don't know. Let's study it together." Sometimes, though, the questions are more along the lines of, "Okay, there are a thousand kinds of bacteria in my house dust. How do I get rid of them all?" The answer is that you shouldn't.

What we ideally want in our homes is a kind of garden. In a garden, you kill the weeds and pests, but you take care of the diverse species you are trying to grow. The species we need to get rid of from our houses are those that can make us really sick or even kill us. But such species are far fewer than you might imagine. Fewer than a hundred species of viruses, bacteria, and protists cause nearly all of the infectious illnesses in the world. As individuals, we keep those species at bay by washing our hands, which interrupts the opportunity for fecal microbes to inadvertently move from feces to hand to mouth. Washing hands doesn't disrupt the thick layer of microbes on your skin; it just removes the most recent arrivals. As individuals, we also keep pathogenic species at bay through vaccination. In turn, our governments and public health systems help to keep the bad species at bay by implementing policies and building infrastructure that provide drinking water devoid of pathogens (but not devoid of life). Our governments and public health systems also help to control the pathogens spread by insects, such as yellow fever and malaria. Finally, doctors should administer antibiotics when (and only when) a bacterial pathogen becomes a problem that can't be controlled in some other way. Together, these approaches to controlling the problem species have saved hundreds of millions of lives and, used appropriately, can continue to do so.

All of these measures work best, though, when they target problem species. When they inadvertently also kill off other species (the other 79,950 or so kinds of bacteria in homes, for instance), the consequences tend to be negative. I'll return often in this book to the question of just what happens when we try to get rid of all of the biodiversity in our houses. Suffice it to say for now that when we do, it tends to make it easier for pathogens to spread, succeed, and evolve, easier for dangerous pests to spread and succeed, and harder for our immune systems to function normally. In the vast majority of cases, it is actually healthier to have *more* biological diversity, particularly the wild biodiversity of soils and forests, in your home rather than less, so long as the dangerous species are under control. It isn't quite that simple (nothing in biology ever is), but nearly so.[25]

This is where some people think to themselves, "I'm still going to try to kill all of it." One of the benefits of the nonpathogenic microbes on our bodies and in our homes is that they help to fight off the pathogens. But you might imagine that if you kill all of the bacteria in your home, there will no longer be any pathogens, nothing the microbes need to fight on your behalf, a clean slate. Cleaning products often advertise that they kill 99 percent of germs (leaving only the truly tough and problematic behind), but maybe you could get that last 1 percent. If there is any home in which this has really been attempted, an indoor space that might offer an example of what is possible, it is the International Space Station (ISS). If you imagine you might scrub your home entirely of bacterial life, the ISS is a perfect illustration of what you will achieve.

Early in its history the National Aeronautics and Space Administration decided it was important to prevent the transport of microbes into space. Initially, the concern was that space shuttles might inadvertently seed the solar systems with Earthly microbes[26] or seed Earth with extraterrestrial life. These remain the primary concerns of NASA's Planetary Protection Office. But with time, NASA scientists also became worried about the potential for astronauts on space shuttles and later the ISS to be trapped for extended periods alongside pathogens. Space itself worked in NASA's favor. The potential for chance colonization of any additional life from space in the space shuttles or the ISS was nonexistent. Open a window in a home on Earth, and outside microbes blow in. Open the hatch on the ISS, and the vacuum of space sucks you (and any life around you) out. In addition, the total volume of air inside the ISS is relatively small, compared, for example, to an apartment building, such that the potential for controlling humidity and the flow of air is, relatively, great. Finally, NASA was able to build a state-of-the-art facility where each bit of food and material headed to the ISS could be cleaned before transport. In short, you are unlikely to get your house to be any more lifeless than is the ISS. The question, then, is whether anything but humans lives on the ISS.

The life on the ISS has been studied in detail and more studies are ongoing. Recently, a new study even searched for life on the ISS using the same approaches we used to study homes in Raleigh. This isn't chance. In 2013, not long after our study on the forty homes was published, Jonathan Eisen, a microbiologist at the University of California, Davis, wrote to me, asking if he could use our protocol to sample the ISS. Much as we had invited participants to sample their own houses, he would invite the astronauts to sample theirs. The same swabs would be used. Similar sites would be swabbed, though there would have to be a few alterations. We had asked participants to swab the dust on their door frames to measure the airborne life that settles around homes. In the low-gravity environment of the ISS, dust doesn't settle. Instead of door frames, the astronauts swabbed air filters. The study also used consent forms similar to those we had used (which gave permission for scientists to examine the data), but with an exception. In studies in homes on Earth, we keep the results of our sampling anonymous (people can see their own results, but no one else can see their results). On the ISS, this wasn't possible. Astronauts are many things but anonymous is seldom one of them. The people living on the ISS at the time were NASA astronauts Steve Swanson and Rick Mastracchio, cosmonauts Oleg Artemyev, Alexander Skvortsov, and Mikhail Tyurin, and the commander, Koichi Wakata, from the Japan Aerospace Exploration Agency. Koichi Wakata swabbed the ISS. The swabs were then brought back to Earth and taken to Jonathan's lab at the University of California, Davis, where they would be studied by Jonathan's student, Jenna Lang.

Earlier studies of the ISS had revealed environmental bacteria to be largely absent on board. Wild species of forests and grasslands were absent. Food-associated species were also absent. If the goal was to get rid of life on the ISS, these were successes. The ISS is not free of bacteria, though. It actually abounds in bacterial life. Nearly all that life just happens to be of one basic kind: bacteria associated with the bodies of the astronauts. This was a key finding of the early studies of the ISS. It was true, too, in Lang's study. To really bring this point home, and give it some context, we can map the ISS and its bacteria relative to other habitats, particularly the forty homes in Raleigh. On this map, samples that are similar in terms of the kinds of bacteria they contain are placed closer together. Samples that are less similar are spaced farther apart. On this map you can see some of what I've already told you about the homes in Raleigh: samples from door sills tend to include both indoor and outdoor species and tend to be similar to each other. Samples from kitchens tend to clump because they contain food-associated bacteria. Also, samples from pillowcases and toilet seats are different from each other, but perhaps not as different as you might hope. The ISS samples are at the bottom of the map, all of them, regardless of which site in the ISS they came from. To the extent that they match anything on Earth, they are like pillowcases or toilet seats.[27]

Like pillowcases and toilet seats, the ISS samples contained fecal microbes. Lang found species related to *Escherichia coli* and *Enterobacter*.[28] She also found a kind of fecal bacterium that has been poorly studied back on Earth, so poorly studied that it doesn't have a name. For now it is called "Unclassified Rikenellaceae/S24-7." The ISS samples weren't identical to those of toilet seats or pillowcases; they tended to have fewer species of bacteria associated with saliva than do pillows, for example, and more of the bacteria associated with skin than do toilets. Previous studies found that stinky-feet bacteria of the species *Bacillus subtilis* are very common on the ISS. Lang found such bacteria to be present, but she found even more bacteria of the genus *Corynebacterium*. *Corynebacterium* species are responsible for armpit odor. What with its *Bacillus* and *Corynebacterium*, it is perhaps not surprising that the ISS has been described as smelling of a mix of plastics, garbage, and body odor.[29] On Earth, we tend to find more *Corynebacterium* armpit bacteria in houses where men live. At the time, the ISS was a house with just men in it. This brings me to something else that was different between the ISS and houses on Earth, namely, the relative rarity of vaginal bacteria, or, rather, the kinds of bacteria that tend to be common in vaginal communities, such as *Lactobacillus* species. One might attribute this reality to the absence of women on the ISS at the time it was sampled.

In nearly every way, the bacteria of the ISS are the sorts of bacteria we would expect in a house on Earth if all the environmental influences were removed. ISS is what you get when you scrub and scrub and close the windows, doors, and hatches. But there was something more. The samples from different sites on the ISS were all very similar to each other. Everything was everywhere. In this one regard, the ISS is like small, traditional homes made of mud or leaves. In such homes, everything is also (compared to other houses) everywhere. But with a difference. In the small traditional houses, be they in Namibia or the Amazon, microbes throughout the home tend to be relatively similar because of the ubiquity of environmental microbes. Environmental microbes are everywhere. In the ISS, Lang found that different sampling sites were similar, but it was because they were all just covered in human bacteria, human bacteria spread evenly in the relative absence of gravity, the absence of gravity and the absence of the rest of life. If you scrub and scrub your home, this is what you may achieve too. It is not unlike what we see in some apartments in Manhattan. And as we and others have begun to study such apartments,

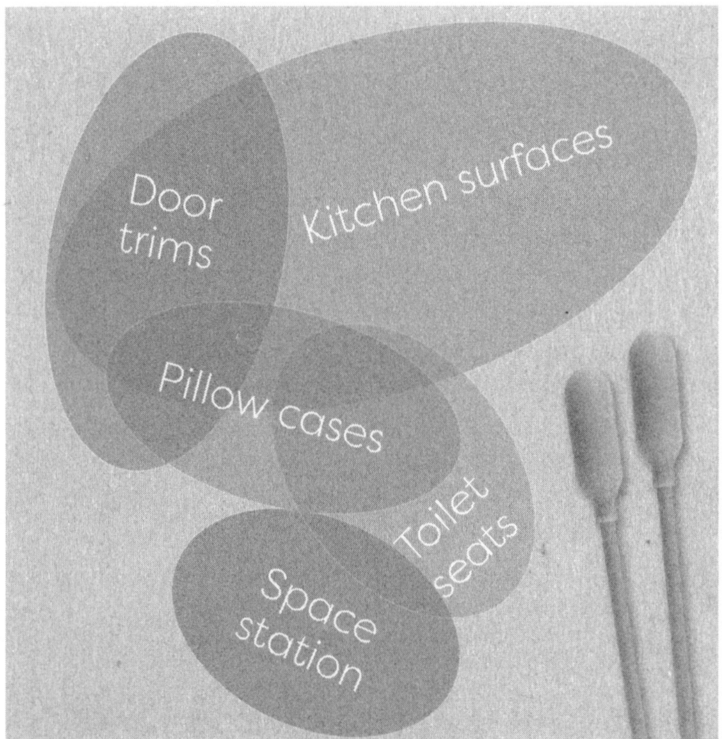


**Figure 3.2** Shapes represent different habitats that have been sampled for bacteria in our study of Raleigh homes and in a recent study of the International Space Station (ISS). The larger the shape, the more the bacterial composition of a particular habitat varies from one sample to the next. The closer together two shapes are, the more similar they are in their composition of bacteria. Habitats at the bottom of the figure tend to be dominated by bodily bacteria, those at the top right by food-associated bacteria, those at the top left by soil and other environmental bacteria. *(Figure by Neil McCoy.)*

we have found a problem. The problem is not what is present but instead what is absent. The problem has to do with what happens when we create homes devoid of nearly all biodiversity except that which falls from us and then, for twenty-three hours of the day, we don't go outside.

# 4

# ABSENCE AS A DISEASE

> In every street the pipes gushed out where decaying rat carcasses drank everything in. . . . Bellies in the air, they floated amid apple peels, asparagus stalks and cabbage cores. . . . it was like a vast infection of tooth decay, like the flatulence of a rotting stomach, like the emanations of a man who has drunk too much, like the dried sweat of rotting animals, like the sour poison of a bedpan. . . . this avalanche of excretions tumbling down the length of the purulent streets . . . let off its nocturnal fragrances.
>
> —*LE FIGARO*

IN THE 1800s the world was seized by spasms of cholera. The first pandemic began in India in 1816 and spread through China, where more than a hundred thousand died. A second pandemic began in 1829 and spread across Europe, and when it was done thirty years later, many hundreds of thousands of people, from Russia to New York, had died. Then, in 1854, cholera started to spread anew, this time globally. In one city after another, whole families died and their dead bodies were loaded into carts together. More than a million people perished in Russia alone. Apartment buildings went from jubilant scenes of the daily circus of work and family to empty shells. In some cities more people died than were born. Ecologists call scenarios in which populations are sustained by immigration alone *population sinks*, a euphemistic term.[1] Cities were sinks down the drains of which human lives poured.

The spread of cholera was blamed on miasma. The miasma theory claimed that diseases, including cholera, were caused by bad-smelling air (the miasma), particularly bad-smelling night air. Miasma is an easy concept to mock, and yet a reasonable sentiment. It encapsulates the idea that bad odors are often associated with illness. Evolutionary biologists argue that an understanding of the relationship between putrid odors and disease is ancient and wired into our subconscious brains.[2] Over our long evolutionary history, avoiding odors we perceived as disgusting would have made our ancestors more likely to survive.[3] Avoiding the smell of dead bodies would reduce the risk of contagion from pathogens on those dead bodies. Avoiding the smell of feces would reduce the risk of getting sick from pathogens in the feces. In this way, the concept of miasma is perhaps so old as to be almost innate. Unfortunately, as cities evolved, the correlation between the odors of putrefaction and disease was no longer useful. Everything smelled bad; to run from the smell was to get out of town, a solution only the rich could afford.

The quest to understand the true cause of cholera was characterized by decades of false starts and a general inability of scientists and the public to pay enough attention to the data in front of them. But in mid-nineteenth-century London, one man, John Snow, was paying a little more attention than were others. Snow had come to believe that cholera was caused by some sort of "germ" passed not from the air but instead from the feces of one person to the mouth of another person. Although feces smelled, the germs themselves, he reasoned, did not. People didn't like this idea. It was at odds with the miasma theory. It was also gross. Then, in 1854, building on work of Reverend Henry Whitehead, Snow collected data on where people were getting sick with cholera and where they weren't in the Soho District of London. Soho had been particularly hard hit by the disease.

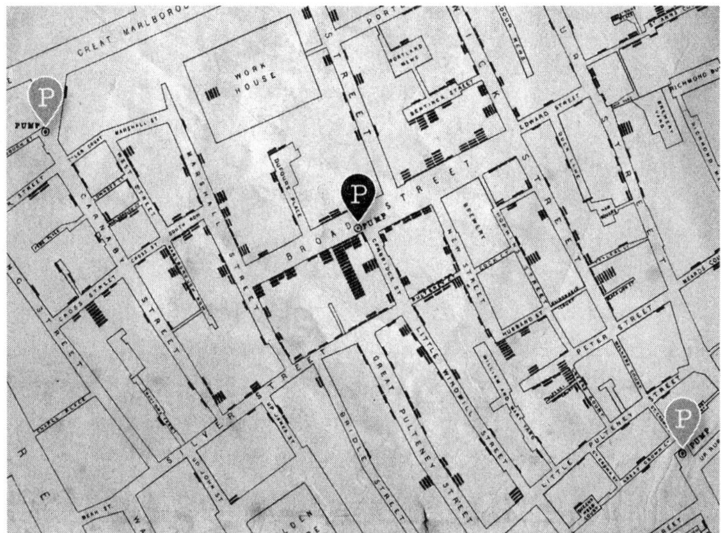


**Figure 4.1** A re-creation of the map on which Dr. John Snow marked the location of deaths from cholera in Soho, London, in 1854. Each black bar marks a death and each ***P*** indicates the location of a water pump. With this map, Snow demonstrated that most of those who died lived near or drank from the Broad Street well. *(Figure modified from map re-created by John Mackenzie [2010], based on original map by John Snow [1854].)*

What Snow eventually came to see was that deaths in Soho were grouped together in a single large, lumpy patch. He discovered why. All of the people in that patch used water from the same well, a well on Broad Street (now Broadwick Street). Some families not using the Broad Street well also died, but as it turned out, those families had also drunk at least a little water from the Broad Street well when their own well smelled of miasma. Snow then mapped the recent deaths from cholera in Soho with the aim of showing that they emanated from the Broad Street well.

Snow used his map to argue that the contamination of the Broad Street well was making people sick and that if the well handle was removed (hence closing down the well), the disease would stop killing the people of Soho.[4] He was right, though it would take years to convince many of his peers of as much. Meanwhile, the cholera epidemic in Soho subsided on its own.[5] It was later shown that the well had been contaminated by an old diaper in an abandoned cesspit next to the well. Years later, Robert Koch, the microbiologist who identified *Mycobacterium tuberculosis* as the cause of tuberculosis, identified the organism that caused cholera, *Vibrio cholerae*. The pathogen had evolved in India and then spread with trade to London and around the world in the early 1800s.

It took decades to figure out how to reconstruct cities so as to deter such contamination; in London the more immediate answer was to pump water into cities from far enough away that it was less likely to be contaminated. After Snow's discovery, cities, including London, began to more actively control disposal of human waste. In some, but not all, cases, they also began to treat the incoming drinking water. Hundreds of millions and perhaps even billions of lives were saved.[6] Interrupting the movement of pathogens from one person's feces to another person's mouth worked.

Inspired by Snow, maps of the spread of disease became a common feature of the field of epidemiology. Students learn that Snow's map was the first illustration of the spread of a disease (not really true). They also learn the power of a map to show the likely origin of a disease and to imply a potential cause. Typically, when maps are used in epidemiology, the goal is to describe when and where a particular species, a pathogen, is present, and then to infer why. Maps depict correlations, but they help epidemiologists think about causation, about how and why. But maps can also betray our ignorance, which happened in the 1950s when a new set of diseases emerged.

Crohn's disease, inflammatory bowel disease, asthma, allergies, and even multiple sclerosis were among this new tribe—the ugly horsemen of malaise and malfunction. All of these diseases were associated with one or another form of chronic inflammation. But what was causing the inflammation?

These diseases were too new to be solely genetic. What was more, just like cholera in London, these diseases had a geography. It was an unusual geography. These diseases, unlike cholera, were more common in regions with good public health systems and infrastructure. The more affluent a region, the more likely it seemed its population was to suffer these diseases. This pattern defied the understanding of "germs" and their geography we'd had since the work of Snow. Yet, in looking at maps of these diseases, in regard to geography or other factors, one might still approach them the way Snow did. He would have used the available maps of the diseases to come up with hypotheses on causes. He would then have looked for natural experiments as a way to test his hypotheses. Finally, once the best hypothesis was tested to his satisfaction, he would have used maps once again to show what he thought to be true. And then, and only then, the details of the biology of the actual agent of disease could begin to be understood. So it was to be with these new diseases. To begin, someone needed to come up with hypotheses and then, on the basis of natural experiments, test them.

The diseases were blamed on new pathogens, on refrigerators, and even on toothpaste. One ecologist, Ilkka Hanski, came to be part of a team that would argue something else entirely was to blame—not exposures to some particular bacterium but instead the absence of exposures. Hanski is an unlikely character in a story about chronic disease and bacteria. He started his career as a world expert on dung beetles. One can read his story, chapter by chapter, in his autobiography. He started to document his life in 2014, writing in a hurry because, as he told his friends in March of 2014, he was dying of cancer. He wanted to get down on paper, for posterity, those things he thought to be most important about the biological world.

In the book, the reader follows Hanski through the stages of his career. Throughout these stages, Hanski was drawn to the study of small patches of island-like habitat. At first those patches were dung piles. For a beetle, a pile of dung is an island that must be discovered and, very rapidly, colonized. Hanski used his own feces or dead fish as bait for the beetles. He attracted and trapped them while hiking up and down Mount Mulu in Borneo to understand the general rules governing when numerous species compete for a pile of shit and when few do. Hanski then moved on to the study of a single butterfly species, the Glanville fritillary (*Melitaea cinxia*), in the Åland Islands off southern Finland. He used this butterfly species to understand ways rare species wax and wane in small patches of habitat. For decades he followed this butterfly and its parasites and pathogens across more than four thousand patches (they are still being tracked). Through this work, he elaborated mathematical models that quantified just how small and isolated habitats could become before the species in those habitats went extinct. Later, Hanski became interested in why some individuals of a particular butterfly species were able to thrive in the face of habitat fragmentation. He discovered versions of genes that seemed to be associated with the ability of some but not other butterfly individuals to live successfully even in a world of small patches of good habitat. Collectively, Hanski's insights, based on fieldwork, theories, predictions, and tests won him the Crafoord Prize in Bioscience in 2011, ecology's version of the Nobel Prize.

Over decades, Hanski's work became ever more focused, moving from study of whole dung beetle communities to an individual butterfly species, to a genetic variant of a butterfly species. Then, suddenly, he began to study chronic inflammatory diseases in humans. A chance meeting inspired the change. In 2010 Hanski saw Tari Haahtela, Finland's preeminent epidemiologist, give a presentation on chronic inflammatory diseases.[7] The material Haahtela covered in his talk was different from anything Hanski had ever worked on or even really seen. It was raw and affecting. Haahtela described the rise in the incidence of chronic inflammatory diseases. He showed that these diseases had all become about twice as common every two decades since 1950; more so in wealthier countries. This rise continues. In the last twenty years in the United States, for example, allergies have increased by 50 percent, and asthma has increased by a third. And as poorer countries

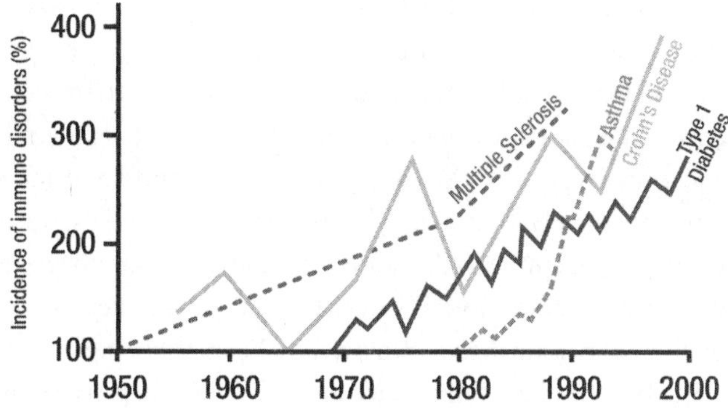


**Figure 4.2** Incidence of immune disorders increased steadily between 1950 and 2000. They continue to increase. *(Figure modified from original by Jean-Francois Bach, published in* New England Journal of Medicine *347 [2002].)*

invested more in urban development, they also saw increases in inflammatory diseases. This global pattern was both striking and worrisome. The rising lines on Haahtela's graphs could be stock prices, human population sizes, or the cost of butter, but for their labels. Their labels bespoke beasts, terrible chronic diseases, stalking the indoors. Haahtela showed maps of places where these diseases were common and where they were not.

Haahtela argued that the illnesses were not caused by pathogens. It was not about germ theory but instead almost its opposite. Haahtela thought that people were getting sick because of their failure to be exposed to species they needed. He didn't know which species those might be any more than Snow knew which contaminant in the well caused cholera. In looking at the maps, Hanski had an idea about what might be missing. To Hanski, the maps and trends that Haahtela showed looked like the inverse of maps and trends he would go on to show in his own presentation on the global loss of old-growth forests and their biodiversity of dung beetles, butterflies, birds, and all the rest. As biodiversity declined through time, chronic diseases seemed to become more common. What was more, in the developed regions where the most biodiversity had already been lost (particularly from people's daily, indoor lives), the diseases were the most common. Hanski thought that perhaps what was missing from people's lives and what was making them sick was not a species but rather something broader. What was missing was biological diversity itself. What was missing for the first time ever in the history of vertebrates, maybe the history of animals, was wild nature. It was missing from backyards, houses, Manhattan apartments, and the International Space Station (ISS) alike.

At this point, Haahtela had already been thinking about the links between biological diversity and disease, albeit using metaphor as much as hard data. In 2009, he had even written a paper in which he noted that in those places in which the diversity of Finnish butterflies had declined, chronic inflammation was more common. He included pictures in the paper of some of his favorite butterflies—the Elban Heath, De Lesse's Brassy Ringlet, the Two-Tailed Pasha, the Polar Fritillary, Nogel's Hairstreak, and a half dozen other species. When the habitat these species needed started to become too fragmented and rare and they started dying out, humans got sick.[8] The butterflies were indicators of a deeper connection between the wilderness outside and the wilderness in people's homes and the consequences of its absence. A disconnection from pathogens such as cholera, one part of nature, benefits humans. But now, humans had gone too far and disconnected themselves not only from the few species that are real and deadly monsters but also from the rest of biodiversity, including beneficial species.

Haahtela approached Hanski. The two men talked. They had met before. Many years earlier, Haahtela, who photographs butterflies as a hobby, pointed Hanski to the Glanville Fritillary as a species on which to focus his studies. As they reconnected, the two men were reminded that they enjoyed each other's company. Both men loved butterflies, and now they were also joined by a common set of megatrends: the loss of biodiversity, the increasing rates of chronic inflammatory diseases, and the societal shift to indoor living, where biodiversity tended to be even more reduced than it had become outside.[9] If they were right that these trends were related, things were only going to get worse. The threat to biodiversity is growing, and our shift indoors, away from biodiversity, is ever more complete. Haahtela invited Hanski to attend one of his lab meetings, where Hanski also met Leena von Hertzen, a microbiologist and a key partner in what would come next. In the meeting, the excitement was the sort that makes arm hair stand on end. Hanski felt, he would later write in his autobiography, as if he had joined one of the most exciting collaborations of his life. Some major component in the world was, it seemed, about to be worked out.

When Snow proposed that feces in water was spreading something to people that was causing cholera, he didn't specifically understand what was being spread. In the same way, Hanski, Haahtela, and von Hertzen didn't understand what aspect of biodiversity loss was making people sick, but they had some ideas about *how* this loss was making people sick. The possibility of a link between exposure to biodiversity and well-being has been tossed around for decades, both in the context of immune health and more generally. E. O. Wilson argued, in his biophilia hypothesis, that we humans have an innate fondness for biodiversity and our emotional well-being diminishes in its dearth.[10] Roger Ulrich has argued that nature reduces stress; Stephen Kaplan, that exposure to biodiversity increases attention spans.[11] The nature deficit disorder extends these hypotheses to consider the ways in which biodiversity, and nature more generally, can promote learning and the psychological well-being of children.[12] The loss of biodiversity, these theories suggest, causes us to ache emotionally, psychologically, and intellectually. Hanksi and Haahtela were influenced by all of this work but thought there was more going on. They thought the loss of biodiversity was also making our immune systems "ache" and malfunction. In their thinking, the antecedent on which they were most directly building was a hypothesis and series of studies arguing that chronic autoimmune diseases were associated with lives that were too clean, too hygienic. This "hygiene hypothesis" was first proposed by David Strachan, an epidemiologist at Saint George's University of London, in 1989. Strachan argued that our modern cleanliness had removed from our lives necessary exposures.[13] Hanski and Haatehla thought that the kind of exposure that was missing was an exposure to biodiversity, the rest of life.

Like small governments, human immune systems are composed of many working parts, organized into chains of command and consequence, and governed by rules that are followed most but not all of the time. In chronic inflammatory diseases, two pathways are relevant. The pathway we have understood for a while is that when a substance (an antigen), be it a dust mite protein or a deadly pathogen, is detected by immune cells on the skin, in the gut, or in the lungs, it triggers a chain of signals that decide whether the immune system attacks the antigen both in that instance, by relying on white blood cells such as eosinophils, and in the future. When the attack response is triggered, a cascade of signals is sent from one kind of cell to another until those signals trigger the recruitment of a circus of different kinds of white blood cells and (in some but not all cases) initiate the production of specific immunoglobulin E (IgE) antibodies. The IgE antibodies remember the antigen and bind to it whenever it turns up again. The key here is to know that this pathway detects antigens, decides whether to attack them, and decides whether to make future attacks easy. If it does this right, it enables the immune system to be able to respond to pathogens quickly. If it does it wrong, the immune system attacks the wrong things, and allergies, asthma, and other inflammatory disorders develop. A second pathway works to balance the immune response by both preventing the buildup of white blood cells such as eosinophils and preventing IgE antibodies from responding to whichever antigen has been detected. This separate pathway (which includes its own specific receptors, a range of regulatory compounds, and signaling molecules) keeps the peace when it is needed, which is most of the time. Most antigens aren't dangerous, especially those that occur frequently and that are associated with ordinary environmental exposures or the species that live on the skin, in the lungs, or in the gut; it is the job of this peacekeeper pathway to remind the body of as much. Strachan, and others, suggested that this peacekeeper pathway, the immune system's soothing voice of reason, wasn't being sufficiently stimulated by ordinary daily exposures. What they could not explain was just what was missing in urban childhoods or childhoods that were too "clean," just what kind of absence triggered this lack of regulation. Hanski, Haahtela, and von Hertzen thought that exposure to biodiversity in the environment, in homes, and on the body somehow helped keep the immune system's peacekeeper pathway functioning normally. In the absence of exposure to biodiversity, the immune system was developing IgE antibodies and inflammation to many antigens that weren't actually dangerous, such as bits of dust mites, German cockroaches, fungi, or even the body's own cells. If children weren't exposed to enough wild species, the regulatory pathway didn't do its job. Allergies and asthma developed, and a host of other problems also ensued. Or so they speculated. These ideas, however exciting they might be, needed to be tested.

Conversations about how and where to test these ideas almost invariably came back to one place, modern Finland. In modern Finland a kind of natural experiment has been under way since the end of World War II. Among Finns, the incidence of chronic inflammatory disorders had increased everywhere except in one place: the Russian half of Karelia, a region that had once been part of Finland but that was no longer. Before World War II, the Karelia region, which stood on the border of Finland and Russia, was united under Finnish rule. After the war, the new Finnish-Russian border bisected the region, yielding Russian Karelians and Finnish Karelians, people with a common heritage but distinct futures.

Today, in Russian Karelia, life expectancies are relatively short owing to traffic accidents, alcoholism, smoking, and every possible combination of each of these problems. In Finnish Karelia, all of these causes of death are less common. For the most part, folks in

Russian Karelia were on the bad side of the fence. But the Finnish Karelians are prone to illnesses that Russians are not: chronic inflammatory disorders. Asthma, hay fever, eczema, and rhinitis were (and have remained) three to ten times more common in Finland than in Russia. Hay fever and peanut allergies are nonexistent in Russian Karelia.[14] Finnish Karelia, on the other hand, is a microcosm of the parts of the world where chronic inflammatory diseases have become ever more common. Since the war, each generation in Finnish Karelia has been more likely than preceding generations to suffer from inflammatory disorders. Not their relatives across the border in Russian Karelia.

Haahtela and von Hertzen had spent the better part of a decade comparing the lives of Karelians on both sides of the border in the aptly named "Karelia Project." On the basis of intensive surveys and blood tests for the IgE antibodies associated with particular allergies, they were able to show that the differences in the prevalence in allergies between the two populations were real. More significantly, they had come to believe that the diseases in Finnish Karelia were due to the absence of exposure to environmental microbes.

Russian Karelians live lives very much like those their ancestors lived fifty or a hundred years ago. They live in small, rural houses lacking central air-conditioning and heating systems, have daily exposure to domestic animals, including cattle, and grow most of their own produce in small vegetable gardens. Their drinking water comes from wells drilled into the groundwater below their homes, or it comes from the surface water of nearby Lake Ladoga. The region is still largely forested and biologically diverse. The Finnish Karelians live in a very different environment. They live in much more developed towns and cities, with much less biological diversity. Compared to Russian Karelians, the Finnish Karelians spend much more time inside, in houses that are sealed more tightly from the outdoors. The life they are exposed to is ever more like that of the ISS and ever less like that of a trail through an old and wild woods.

Haahtela and von Hertzen along with their students had shown that some plant-associated microbes seemed to be missing from the daily lives of children growing up in Finnish Karelia, but they hadn't connected all the pieces. Now, with Hanski involved, they started to develop their full argument, which was that the loss of outdoor biological diversity (be it butterflies, plants, or anything else) led to the loss of indoor biological diversity, which then led to immune systems in which eosinophils were too abundant and, as a result, chronic inflammatory diseases. In a paper led by Leena von Hertzen, the scientists called this idea the "biodiversity hypothesis,"[15] which together they would proceed to test.

The ideal project would have been to experimentally alter the biodiversity to which children were exposed in their homes and backyards and then follow those same children over the next few decades. In theory, this might have been possible, but the process would have also been very expensive and prolonged. Another approach would have been to compare the lives and exposures of the Karelians in Russia and in Finland, but this proved untenable at the time. So Hanski, Haahtela, and von Hertzen decided to take a third approach. They would study a single region within Finland, one Haahtela and von Hertzen had been working in since 2003. In this region, they would test whether teens (fourteen- to eighteen-year-olds) living in houses with less biodiversity were more likely to have immune systems prone to allergy and asthma.

The chosen region was a square, 100 kilometers by 100 kilometers. It included a small town, villages of different sizes, and isolated houses. Within the region, Haahtela and von Hertzen randomly selected homes. Nearly all of the families in the randomly selected homes had not moved in many years, hence their teens had grown up in the same house in which they now lived (an impossibility in many regions). One might criticize the scientists for not choosing a more diverse region or working in many regions. One could criticize many things. But as the ecologist Dan Janzen has often noted,[16] the Wright brothers did not take off in a thunderstorm. Hanski, Haahtela, and von Hertzen chose to begin their work where they could control as many extraneous factors as possible and build on the data already in hand.

The team tested each of the teens for allergies. They then also measured the biodiversity in their backyards and the biodiversity on their skin. They predicted that teens with less biodiversity in their backyards would have less biodiversity on their skin and would, in turn, be more likely to have allergies. They measured biodiversity by counting the number of kinds of nonnative plants, native plants, and rare native plants present in backyards. Each plant tends to have its own associated bacteria and fungi and even its own associated insects, so measuring plants was a way to capture a simple proxy of the rest of the life the teens might encounter. Plants are also easier to measure than other organisms because they are visible (in contrast to microbes) and don't move (in contrast to, say, butterflies or birds).[17] Skin bacterial biodiversity was measured on the middle of the forearm of teens' writing hands. Bacterial biodiversity was tallied in much the same way we tallied it in homes in Raleigh. Finally, allergies were measured as a function of the IgE antibodies in the blood of the teens. In general, more IgE equals more allergies. In those teens with high IgE levels, the team also tested for allergies to specific antigens such as cats, dogs, or mugwort.

The study was straightforward, and each investigator had a specific role. Haahtela was in charge of blood samples to test for allergies, von Hertzen was in charge of skin samples to measure bacterial communities, and Hanski was in charge of sampling and studying the diversity of plants. Everyone worked together on the analyses. It was exciting, a potentially big step forward, though also, in some ways, far-fetched.

WHEN HANSKI AND his colleagues looked at the data, they were excited but also anxious. Would the diversity of plants in the homes of the teens really matter? Although the scientists had controlled for as many factors as they could, predicting differences in the health of humans is notoriously difficult. For Hanski, in particular, this was trying. Humans, he was quickly learning, were far more difficult to study than were dung beetles and butterflies. He would have liked to have done an experiment. He worried that if they found no pattern, the study wouldn't mean anything. Perhaps they would just need to study more teens, or more countries, or across more years.

But what Hanski, Haahtela, and von Hertzen saw was, to them, remarkably clear. Teens living in houses with a higher diversity of rare native plants in their backyards had different bacterial species on their skin. They tended to have a higher diversity of bacteria on their skin, particularly those kinds of bacteria associated with soil. Presumably, those bacteria were landing on teens in their backyards or maybe coming into their homes through open windows and doors and landing on them while they went about their days or even slept. In addition, teens with a higher number of rare native plant species in their backyards and more diverse skin bacteria were also at a reduced risk of allergies. Any allergies.[18] The scientists hadn't done an experiment; they had merely observed a correlation, but the correlation they observed was entirely in line with their hypothesis.

In particular, one group of bacteria, the Gammaproteobacteria, seemed to be more diverse when plant diversity was high and more common on teens with fewer allergies. More than forty years earlier, species of this same group of bacteria had been shown to vary in abundance on human skin with the seasons.[19] In the samples Megan Thoemmes took from chimpanzee nests, the abundance of Gammaproteobacteria also varied from season to season. Hanski, Haahtela, and von Hertzen found that the Gammaproteobacteria also varied in space. Again, it didn't matter whether they considered allergies to cats, dogs, horses, birch pollen, timothy grass, or mugwort. In each case, individuals with more kinds of Gammaproteobacteria, particularly more kinds of one genus, *Acinetobacter*, on their bodies were less likely to have allergies. In a subsequent study, Hanski and Haahtela, along with another group of researchers, were able to show that individuals (again in Finland) with more of a kind of *Acinetobacter* on their skin tended to have immune systems that produced more of a compound associated with immunological peacekeeping.[20] This same peacekeeping compound was also produced by mice in the lab when they were experimentally dosed with *Acinetobacter*.[21]

An additional test of the idea that bacterial diversity, and presence of *Acinetobacter* in particular, was helping to keep allergies in check was to compare the bacteria on the skin of teens in the Russian and Finnish parts of Karelia. Haahtela led a separate study to find out. The backyard biodiversity should be higher in Russian Karelia than in Finnish Karelia. It was. The skin biodiversity should be higher in Russian Karelia than in Finnish Karelia. It was. Finally, the abundance of *Acinetobacter* bacteria should be higher on the skin of teens in Russian Karelia than in Finnish Karelia. It was, too.[22]

We can see in Hanski, Haahtela, and von Hertzen's results a direct relationship between exposure to native plant diversity and the effect of native plant diversity on Gammaproteobacteria on the skin (and other bacteria with similar effects in the lungs and gut), which in turn triggers the peacekeeping pathway of the immune system and keeps inflammation in check.[23] We achieved such exposures for tens of millions of years without even having to try. Gammaproteobacteria are diverse in wild plants but also in our food plants. They live as mutualists of seeds, fruits, and stems. We breathed them, we ate them, we walked through them. Then we moved indoors, where the Gammaproteobacteria disappear. They seem to be rare in food plants kept very cold. They disappear when our food plants are processed. They were entirely absent from the ISS and are rare in most urban apartments we have studied. Maybe the diversity of Gammaproteobacteria could be useful not only in the garden but also in potted plants indoors and in fresh fruits and vegetables.[24] To test the specific role of Gammaproteobacteria, scientists would need to alter the diversity of plants in backyards, bring a diversity of plants into homes, feed families fresh fruit and vegetables that have (or have not) been sterilized, and then study whether those changes alter, over years, immune health. It would be analogous to when Snow took off the well handle, just something of the opposite, letting biodiversity flow back in. One could do this. No one has.[25] One study, though, has come close, building on insights from the study of Amish children, Hutterite children, and mice.

The Amish and the Hutterites both moved to the United States in the eighteenth and nineteenth centuries. Genetically, they have similar backgrounds, particularly with regard to genes known to influence susceptibility to asthma. Culturally, they tend to live relatively similar lives: they eat the same German farm foods, have large families, get vaccinated, drink raw cow's milk, and otherwise go about life in remarkably similar manners. Neither group watches television or uses any other sort of electricity. Nor does either group believe in keeping animals as pets. All domestic animals in both communities are working animals. In both groups, marriage outside of the group means leaving the group. At a glance, their genes, lives, and experiences are the same. The main difference, biologically, between the Amish and the Hutterites is that the Hutterites have decided to practice industrial agriculture. They drive tractors. They use pesticides. They plant relatively few varieties of crops. In contrast, the Amish farm as they have always farmed, using horses for labor. Amish children are more directly, physically, connected to their fields, animals, and soils than are Hutterite children. Also, the front doors of Amish houses tend to be fifty feet or so away from the barn door, whereas Hutterite houses and Hutterite farms are often separated by great distances. And, as Hanski, Haahtela, and von Hertzen might have predicted, given this difference, asthma is rare among the Amish. The Hutterites, meanwhile, suffer from asthma at rates higher than almost anyone else in the United States. Twenty-three percent of Hutterite children have asthma. And, just like Finnish kids with few wild plant species in their backyards, the

Hutterite kids have elevated levels of the IgE antibodies against common allergens in their blood. Nor are these differences in IgE antibodies the extent of the immunological differences.

Recently, a large scientific team, headed by scientists and clinicians at the University of Chicago and the University of Arizona, compared the immune systems of Amish children and Hutterite children. When the team from the University of Chicago studied the blood of a sample of Amish and Hutterite kids in more detail, they found that when challenged with a compound associated with the cell walls of bacteria, the blood of the Amish kids produced fewer of the compounds, cytokines, that signal alarm. In addition, the Amish kids had different kinds and quantities of white blood cells. They had fewer eosinophils, the white blood cells most associated with inflammation. Also, their neutrophils tended to be of a variety that was, to put it plainly, less likely to indiscriminately attack. Finally, the Amish kids had more of a variety of monocyte (yet another kind of white blood cell) associated with suppressing the immune system. In short, the blood of the Hutterite kids was a schoolyard thug and, in comparison, the blood of the Amish kids was peaceful.

The team from Chicago and Arizona decided that one way to isolate the effects of the Amish dust and its microbes on immune systems was to experimentally give individuals with inflammatory diseases doses of the dust. They couldn't ethically do this experiment on humans, but they could do it with mice. Scientists have bred a variety of mice that suffers from a chronic inflammatory disease akin to allergic asthma. The mice develop asthma symptoms when exposed to egg proteins. Egg proteins are their Kryptonite. The team gave the asthmatic mice three treatments. One group had egg proteins sprayed into their noses every two or three days for a month. One group had egg proteins plus dust from the bedrooms of Hutterite families sprayed into their noses the same number of times. The third group got egg proteins and Amish bedroom dust (dust that it would later be shown tended to have more kinds of bacteria than the Hutterite dust, or more biodiversity). The mice given the egg protein suffered an allergic response akin to asthma. No surprise. The mice given the egg protein and Hutterite dust actually suffered a worse allergic response than did those that got just the egg. But what about the mice given egg protein and Amish dust? The Amish dust nearly completely prevented the allergic response of the mice to the eggs. Not only did the biodiverse Amish dust keep the mice from getting sick, it damn near made them well, even though they were getting dosed every other day with egg protein, their great weakness.[26] A Finnish team was able to show a similar effect in mice using dust from barns in rural Finland (but not dust from homes in urban Helsinki).[27] This isn't to say that if you are asthmatic you should go sneaking around Amish bedrooms or Finnish backyards snorting dust (especially not without asking permission), but it may well suggest that you need to be sniffing more biodiversity, more of the wild.

The special stuff in the Amish house dust may have been Gammaproteobacteria that triggered the peacekeeping pathway in the lungs (rather than on the skin) in keeping with the prediction of Hanski and colleagues. But even if it isn't, even if what is key in the lungs and gut is some other group of bacteria such as, say, the Firmicutes and Bacteroidetes, or maybe even some special fungi, the researchers' work offers a broader insight that has as much to do with how they framed their question as it does with the details of what they saw: as our exposure to biological diversity, including the biodiversity of plants but also animals and much else, decreases, the odds that we are exposed to the right bacteria, including Gammaproteobacteria, also decreases. We can consider this as probabilistic. Imagine there is a certain number of kinds of bacteria to which you need to be exposed to stay healthy. If this is the case (and given that we don't know where to even find most kinds of bacteria), the more plants and animals and soil you interact with, the more likely you will pick up some of those key bacteria. The fewer kinds to which you are exposed, the less likely you get the right ones, the ones that activate your innate immune system in the right way to keep the eosinophils in check. But chance is chance, so you could also be exposed to great biodiversity and fail to get what you need; some Amish kids get allergies just as do some kids who live in the Russian part of Karelia; it is just less likely.

It would, of course, be much more satisfying to figure out exactly which of these bacteria we need, make sure we are exposed to them, and leave it at that. Until we do, we are just one step beyond the miasma stage of understanding chronic inflammatory diseases. Taking another step may take a while. Consider the fecal transplant. The best treatment for people whose gut ecosystems are invaded by the weedy pathogen *Clostridium difficile* is a fecal transplant. In a fecal transplant, a sick person is given a heavy dose of antibiotics. The feces and fecal microbes of a healthy person are then transplanted into the sick person as an attempt to restore the sick person's ecosystem. It works. Many lives are saved thanks to fecal transplants, which restore enough of the gut ecosystem to prevent *Clostridium difficile* from thriving. For practitioners, fecal transplants have been a great relief for the treatment of patients who had few other options. Microbiologists, too, have hailed them as innovative, a sign of the future. But they are also an acknowledgment that we don't know which species is essential, and in lieu of more knowledge, the best option is to restore everything, to reboot and rewild the gut.

Scientists love making and testing predictions. Among the most predictable features of science is its sociopolitics. I predict that over the next ten years any of a variety of pills and treatments will be offered up as ways to cure yourself of chronic inflammatory problems. Some scientists will continue to suggest that the key missing factor is our exposure to particular kinds of tapeworms, hookworms, and other wrigglers. Others will suggest it is exposure to Gammaproteobacteria. Others, that it is exposure to a single species of bacteria, though different labs will argue for different species, and some will argue that we need those bacteria in our food, others that they are more necessary in our water. In the meantime, someone will find a set of human genes that seems to make some people more susceptible to these diseases than others. It will become apparent that different people need exposure to different microbes based on their genetic backgrounds. But, in those studies, the geneticists will realize (relatively late) that they have mostly sampled white male college students and that when they consider a truly diverse population, the story will become even more complex. Ultimately, it will appear that the microbes, or at least exposures to microbes, people need to stay healthy depend on where people live and even their culture. Maybe a perfectly prescriptive model will come out of this suggesting what each person should do. I wouldn't bet on it, but we need to continue to try to figure it out. It was useful when Snow figured out the mode of transmission of cholera. It was even better when the cholera bacterium, *Vibrio cholerae*, was identified and water systems could be tested for its presence to be make sure they were safe to drink.

While we wait for perfect clarity, we can acknowledge the problems with the status quo and choose an alternative approach, one that is not perfect and yet that is undeniably better. The status quo is that we are exposed to far different species than we used to be, far fewer because we have diminished the biological diversity of the world around us and because we spend nearly all of our time indoors, a realm we appear to be making ever less diverse. As a result, Crohn's disease, asthma, allergy, multiple sclerosis, and their kin have become far more common. What then can we offer our children? We need to offer them the chance to interact with a diversity of microbes that, in doing so, increases the odds that they are exposed to the ones they need. Play the ecological lottery more times and you better your chances of getting it right.

Plant a greater diversity of plants outside of your home and interact with those plants. Tend them. Watch them. Take a nap on them. It may well be that having a greater diversity of plants indoors triggers the same sorts of benefits. Grow a garden and sink your fingers into the soil. Or go full Amish and get a cow that you keep near your backdoor. This may well help and it won't hurt. Meanwhile, we also need to make sure that whatever species we most need still exist in the future. We need to, as Haahtela put it in 2009, "take care of the butterflies," which is to say, save the biodiversity of life around in general until we know for sure what we need. Save the butterflies for our own good. Save the butterflies because where butterflies are diverse and wild, so too are the microbes, so too are species that we may need but have yet to study. Save the butterflies to pay homage to Ilkka Hanski. Hanski died on May 10, 2016, still in love with butterflies. He died still fascinated by the workings of the world. He died aware that though the flapping of a butterfly's wings may not change the weather, the extinction of butterflies, or the plants on which both the butterflies and many bacteria depend, can make us sick. We need biodiversity in order to be well. We need it in our backyards and in our homes; we may even need it, as it turns out, in our showerheads.

5

# BATHING IN A STREAM OF LIFE

We must conclude that there are more animalcules or minute fishes in the sea, than has ever yet been thought of.

—ANTONY VAN LEEUWENHOEK

I bathe once a month, whether I need it or not.

—QUEEN ELIZABETH I

In wine there is wisdom, in beer there is freedom, in water there is bacteria.

—SIGN ON THE WALL IN A PUB IN DUMFRIES, SCOTLAND

IN 1654, REMBRANDT painted a woman, in Amsterdam, bathing in a stream. The woman has placed an elegant red robe on a rock. She wades in the water, lifting her nightdress above her knees so that it does not get wet as she treads deeper. It is night and dark and the woman's skin glows as it disappears into the water. The painting evokes earlier works from ancient Rome and Greece. The woman who is stepping into the stream in Rembrandt's painting is stepping from one world into another. Among art historians, the transition she is making is metaphorical.[1] But to a biologist like me, it is also ecological. In entering the water, she is suddenly exposed to an entirely new group of species, of microbes, of fish, and much else. We imagine water to be clean and we imagine clean to mean lifeless, and yet all water you have ever bathed in, swam through, or drunk has been full of life.

The stream Rembrandt painted looks like the water in the small canals and streams near Amsterdam. The woman is probably Rembrandt's lover, Hendrickje Stoffels. Even if Rembrandt did not intend to depict a particular body of water, his reference and inspiration would have been what he knew and had seen. Assuming this water was similar to the water of nearby Delft a decade or so later, it is likely to have been inhabited by microbes very much like those that Leeuwenhoek described living in the canal in front of his house. Of course, the water to which you expose yourself today is likely very different from that to which Rembrandt's lover exposed herself. By different I don't mean it is devoid of life; I mean only to say that as you slip into the bath or stand beneath the shower, you are covered in life-forms likely to have been rare in Delft, entirely different beasts. Recently, I found myself thinking a lot about these species.

It all started when Noah Fierer, my collaborator at the University of Colorado with whom I first studied the dust in houses, emailed me in the fall of 2014 to say he had an idea for a project. He had stumbled upon a showerhead mystery, and a real doozy at that. "Are you in?" he asked, before explaining what I might be in for. "I've been talking to people about showerheads and we need to work on them. It is going to be great." What followed was the kind of shorthand conversation upon which discovery depends, a conversation in which Noah provided a brief sketch of an idea and then, having offered that sketch, assumed that, based on the sketch, I would fill in the blanks. "Cool either way," he said, which is code for, "If you aren't in, you are missing out and I won't let you forget it, ever, but you don't have to collaborate if you don't want to. But if you are in, let's get it moving."[2]

The sketch was this. The tap water that flows into your house, and eventually into and through your showerhead, is alive. Leeuwenhoek found both bacteria and protists in rainwater as well as in the water from his well. So too have subsequent researchers. Tap water is no more or less full of life than is rainwater. In Denmark, where I work for part of the year, for instance, one can encounter small crustaceans in tap water.[3] In Raleigh, where I can be found the rest of the year, tap water contains the bacterial species *Delftia acidovorans* with reasonable frequency.[4] *Delftia*, which was first isolated in the soil of Leeuwenhoek's own Delft, has the ability to concentrate minute quantities of gold found in water and precipitate them. It also has unique genes that allow it to thrive in mouthwash (or a mouth that has recently been mouth-washed). All of this has long been known and is interesting, but neither news nor Noah's focus. What had intrigued Noah was the knowledge that as water passes through pipes in general and showerheads in particular, a thick biofilm builds up. *Biofilm* is a fancy word that scientists use to avoid saying "gunk."

Biofilm is made by individuals of one or more species of bacteria working together to achieve the common goal of protecting themselves from hostile conditions (including the flow of water, which constantly threatens to wash them away). The bacteria make the infrastructure of the biofilm out of their own excretions.[5] In essence, by working together, the bacteria poop a little indestructible condominium in your pipes, a condominium built of hard-to-breakdown complex carbohydrates. Noah wanted to study the species living in the biofilm of showerheads, species fed by tap water and, when the pressure is high enough, let loose into the fine aerosol spray of water droplets pelting our hair and bodies and splashing up and into our noses and mouths.[6] He wanted to study them because they were interesting but also because, in some regions, but not others, they increasingly seem to be making people sick.

The bacteria in biofilms making people sick are species of the genus *Mycobacterium*. Mycobacteria are different from most waterborne pathogens, such as *Vibrio cholerae*. The normal habitat of the *Mycobacterium* species found in tap water is not the human body. Instead, they live in the pipes themselves. These pipe-loving mycobacteria (plumbingophiles) are not ordinarily pathogens. They become problematic only when they, quite accidentally (from the perspective of their own well-being), make their way into human lungs. In this, mycobacteria and several other pathogens associated with the new habitats we have made in houses (such as *Legionella* bacteria) represent a challenge very different from those we normally encounter when considering pathogens, a challenge associated with the ways we build our homes and cities.

The *Mycobacterium* species in showerheads are typically referred to as NTM, where *NT* stands for "nontuberculous" and M stands for "mycobacteria." This means, as you may have inferred, other mycobacterial species are tuberculous, namely, the species *Mycobacterium tuberculosis* and its close relatives. We imagine the worst monsters of history to be beasts with bad breath and extra arms that one battles using a shield and a longsword, the beasts out of, say, Viking sagas. But the real demons of the past looked far more like *Mycobacterium tuberculosis*. Invisible to the naked eye, they "looked" like nothing other than their consequences, which were horrible deaths.

*Mycobacterium tuberculosis* is the cause of tuberculosis in humans. Tuberculosis killed one in five adults between 1600 and 1800 in Europe and North America.[7] *Mycobacterium tuberculosis* appears to have long associated with humans and our extinct relatives and ancestors. The dangerous form of the pathogen evolved at about the time modern humans moved out of Africa (at about the same time that we have the first good evidence of houses and we started coughing on each other more). *Mycobacterium tuberculosis* spread with us. Once we domesticated goats and cows, we then gave them *Mycobacterium tuberculosis*, and, in their very different bodies, confronted with the realities of their unique immune systems, *Mycobacterium tuberculosis* evolved into *Mycobacterium caprae* in goats and *Mycobacterium bovis* in cows. We gave *Mycobacterium tuberculosis* to mice and it evolved into a form better able to take advantage of the immune systems of mice. We gave it to seals, in which it evolved yet another form. That form appears to have traveled, with seals, to the Americas no later than 700 CE, where it infected Native Americans (and then evolved into yet another specialized form).[8]

In each case, the bacteria rapidly evolved special traits enabling them to better survive and spread among individuals of each new host. The immune system and body of a seal are different from those of a human and so require special tricks. So too the immune systems and bodies of mice, or goats, or cows. Individual lineages of the microbe evolved these tricks. The human form of the pathogen even appears to be adapted to different populations of human hosts (which, because tuberculosis is deadly to even relatively young people, also led to adaptations of those human populations to the pathogen). *Mycobacterium tuberculosis* is an emblematic example of evolution's mechanisms every bit as elegant as that offered by the differences in beak shape among the species of Darwin's finches.

Antibiotics, first developed in the 1940s, allowed us to gain a real victory against *Mycobacterium tuberculosis*, but today, many strains of tuberculosis bacteria are resistant to most antibiotics. Our shining weapons, once the great silver bullets of medicine, now seem ever more like wooden swords. Resistant strains are (predictably) spreading. All of this is to suggest that the lineage of mycobacteria is one about which it would be good to have a robust awareness. Nothing prevents the nontuberculous mycobacterial species found in showerheads from adapting, as *Mycobacterium tuberculosis* has, to take advantage of us. They might adapt to better thrive in our water systems or, more worryingly, our bodies.

So far, the risk of infections due to nontuberculous mycobacteria is high only for immunocompromised people, people whose lungs have an unusual architecture, and individuals with cystic fibrosis. In these individuals, nontuberculous mycobacteria can cause symptoms like those of pneumonia, as well as skin and eye infections. Unfortunately, the risk of nontuberculous mycobacteria infections is increasing overall in the United States, but just how common infections are and how much more common they are becoming vary geographically. Some regions seem to have many more infections than do others. In some regions, such as California and Florida, infections are common. In others, such as Michigan, they are rare. This difference could be due to differences in either the abundance or the presence of mycobacterial species in various regions. The species in Florida, for example, don't seem to be the same as those in Ohio, and this might matter.[9] Also, the mycobacterial species associated with infections tend to be the same species and strains found in showerheads, which are different from those associated with soil or other wild habitats.[10]

On the basis of the information I just gave you about mycobacteria, I could more or less guess what Noah had in mind for our impending showerhead investigation, the method for our plunge into the gunk. I could guess because in the years since we first studied forty houses in Raleigh, Noah and I have developed a way of working together that repeats itself. And, anyway, he had me at the words "showerhead mystery." I responded to Noah's email, committing, in a sentence or two, to coordinating the sampling of showerheads around the world.[11] So began what is probably the largest ever study of the ecology of showers and showerheads. It is based on trust: I trust that nine times out of ten, if Noah is excited about something, then it is probably something pretty interesting.[12] I've never heard anyone talk about trust in science, yet it influences what I do in the lab every day. A huge component of modern science is social, and inside a researcher's most trusted social group, that quorum of very trusted colleagues, science moves faster. Everything happens faster. Conversely, most scientists have colleagues they don't trust or colleagues with whom trust hasn't yet been established. Such collaborations are slower, more deliberate, less able to respond to wild schemes in the middle of the night. I trust Noah, so, with him, I'm game for a wild scheme. We have now worked together on a half dozen major projects (beetle armpits, belly buttons, the microbes of forty houses, the microbes of a

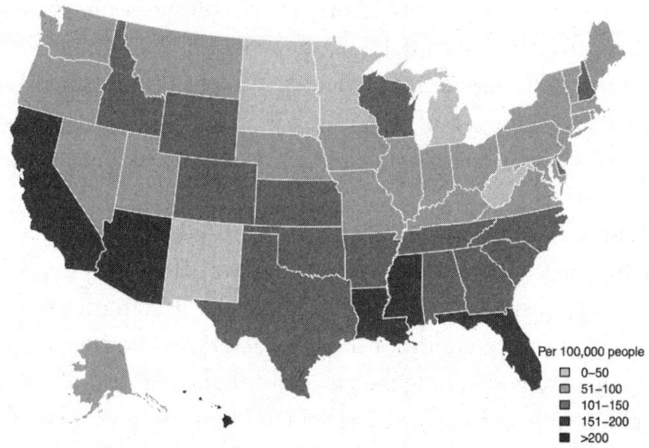


**Figure 5.1** Map showing the prevalence of pulmonary nontuberculous mycobacteria cases in the United States between 1997 and 2007 among a sample of adults aged sixty-five and older. Hawaii, Florida, and Louisiana are among the states in which mycobacterial infections are most common per capita. Much as when Snow mapped cholera cases, for us, a key piece of solving the mycobacteria mystery is mapping where both NTM infections and *Mycobacterium* species occur. *(Data from J. Adjemian, K. N. Olivier, A. E. Seitz, S. M. Holland, and D. R. Prevots, "Prevalence of Nontuberculous Mycobacterial Lung Disease in U.S. Medicare Beneficiaries,"* American Journal of Respiratory and Critical Care Medicine *185 [2012]: 881–886.)*

thousand houses, global forensics, and more). The science we work on together comes easily (albeit, as that list suggests, at times also idiosyncratically).

Earlier in 2014, I had just finished collecting data for a project in which my Danish colleagues and I engaged children in Danish schools to take samples of the life flowing from the water fountains and spigots of their schools. So I knew something about the life in water, but in considering the special case of showerheads, there was more to learn. In Denmark, we found thousands of kinds of bacteria in tap water, as have similar studies of tap water in the United States and elsewhere in the world. The species we or others have found in tap water include bacteria, amoebae, nematodes, and even small, leggy crustaceans. Although tap water is high in biodiversity, it is typically low in biomass, the living mass of life. Tap water does not contain much that might be construed (even by bacteria) as food. Nutritionally, it is a kind of liquid desert, such that many species persist, but none thrive. The biofilms in showerheads are different.

The water that flows through showerheads tends to be warm, which makes it easier for bacteria to grow. It also tends to pool for many hours between uses (which keeps the bacteria from drying out). Given these conditions, once bacteria and other microbes establish in biofilms along the pipes within showerheads, they have the environment they need. In that environment, like sea sponges, they are able to harvest whatever flows past them. The more the water flows, the more they can harvest. The resources available in any particular drop of water are modest, but the resources available in the gallons and gallons of tap water that flow collectively through a showerhead can actually be quite great. As a result, the biomass in showerheads is twice or more that in the tap water itself. What's more, that biomass is composed of far fewer species than in the tap water, hundreds or even just tens rather than thousands.[13] These species come to form relatively stable ecosystems in which each species plays a role. In biofilms one can even find predatory bacteria swimming, as Leeuwenhoek might have said, like "pike[s] through water." Right now, in your showerhead, these tiny "pikes" are latching on to other bacteria, drilling holes in their sides, and releasing chemicals that digest them. Showerhead biofilms also sustain protists that eat the "pikes," and even nematodes that eat the protists, as well as fungi doing their own fungal thing. This is the food web that falls upon you as you bathe. Each day, life falls on you, midmeal (theirs, not yours; though that too), flipping back and forth, stunned by the disruption.

In the average American showerhead, the biofilm that grows contains many trillions of individual organisms, layered as much as half a millimeter thick. The mystery was why these showerheads sometimes abound in mycobacteria and in other cases lack them entirely. When we began our project, no one could explain these differences. In considering an ecological system like a showerhead, about which little is known, my intuition about the first step is nearly always the same. My intuition, like that of any scientist, reflects some mix of my scientific training, what I am good at, and what I enjoy. What I always want to know first is how life with all of its attributes (its abundance, diversity, and even consequences) varies from one region to the next. In the context of showerheads, I wanted to know how many species the most diverse showerheads had, where showerheads were most diverse, and then, for *Mycobacterium*, how the identity and abundance of particular species vary from place to place. To me, until we know about the patterns in such variation, it doesn't make much sense to take any next steps because we don't really know what it is we need to explain (conversely, to some scientists, this step is not even viewed as part of science, which is to say we scientists differ as much, perhaps, as do our showerheads).

Our first step, then, was to engage people around the world to swab their showerheads and, having done so, send us back a sample of the scuz they found. Folks in my lab would then organize the data about the person who took the sample. The sample would then be sent to Noah's lab, where his technicians, or postdoctoral researchers, would decode the sequence bits of the DNA to produce at least a coarse list of the bacteria and protists present in each sample, including mycobacteria and other potentially problematic species such as *Legionella pneumophila*, the cause of Legionnaire's disease. Matt Gebert, one of Noah's students, would then identify the specific species of *Mycobacterium* present in the sample by decoding a specific gene (*hsp65*) known to differ from one *Mycobacterium* species to the next. The samples would then be sent to other collaborators, each of whom would study some other aspect of the story, culturing the microbes from the showerheads, for example, in order to decode their entire genomes, base by base. It was to be an all-taxa biological inventory of the world's showerheads. But first, we had to convince people to send us samples from their showerheads.

We used our social networks to search for people around the world willing to be part of our project. We tweeted. We wrote blog posts. We contacted friends and collaborators. We tweeted again. Many people were interested and signed up. We then got ready to send out kits, but before we could do that, people who had read the protocol started to send us questions about the project. Reaching out to thousands of potential project participants is a good way to quickly learn what you do and do not know about a particular topic and about the clarity of your protocol. Thousands of people simultaneously begin to pay attention to something they had not paid attention to quite so much before. This initial moment of project engagement can be revelatory, though not always in the ways one anticipates. In the context of showerheads, it quickly became clear that we did not know enough about the geography of showerheads themselves. In the American showerheads we worked with initially, we could unscrew the top of the showerhead, look the shower scuz right in the eyes (or where the eyes would be if shower scuz had eyes), and swab it. We had proposed this be done in Europe as well. We hadn't accounted, however, for differences among countries in the type of showerheads people like to use. We began to receive emails from disgruntled Germans indicating that we knew nothing about how a German showers. German bathrooms (and, it would turn out, those of most other Europeans, though it was only the Germans who emailed) have showerheads attached (permanently) to flexible hoses. This precluded the kind of sampling we described in our protocol. The Germans were writing to tell us as much. The emails came to me. They went to various people in my lab. They even, when we didn't answer fast enough, went to the department's administrative assistant, Susan Marschalk. And then, when she didn't respond fast enough (to say she was the wrong person to contact), emails were sent to other people with even less to do with the project. Department head. Deanlet.[14] Frustrated emailers know no bounds. In response, we

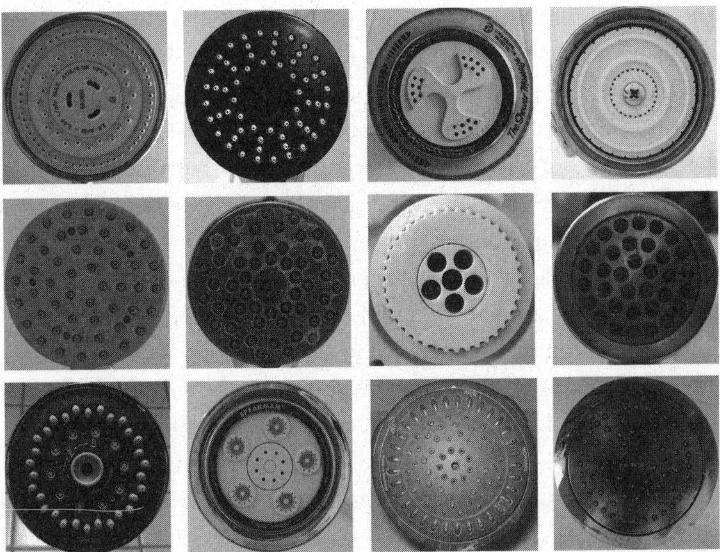

**Figure 5.2** A diversity of showerheads. From this diversity (and more), we sampled microscopic life. Out of the holes in showerheads, be they big or small, sprays a wilderness. *(Showerhead photos by Tom Magliery, flickr.com/mag3737.)*

changed our protocol to accommodate European showerheads. We would soon realize that the hoses were not the only differences between showerheads in the United States and Europe, not hardly.

Showers are a very, very modern contrivance with complex consequences for our bodies, consequences no one anticipated when we first began to stand beneath them. For most of our mammalian history, our ancestors didn't take showers or baths. They probably didn't even really swim very often. They might have cleaned themselves, clumsily. Cats bathe themselves with their tongues. Dogs do the same, albeit less rigorously. But even a moment spent contemplating this possibility (try licking your own lower back) in our own history suggests that it has been a long time since this has been possible for us. Many nonhuman primates groom each other, but the grooming mostly has to do with picking off visible bits and pieces of things that might be lice (or might not be). Similarly, some mammals roll in soil or mud,[15] but this, too, seems to be more about controlling animal parasites such as lice than it does with controlling microbes or odors. Some Japanese macaques bathe in hot springs, but they do so to warm up.[16] Chimpanzees that live in savannahs get into the water occasionally, but only when it is really hot, presumably to cool off. Chimpanzees that live in rain forests don't deign to dip.[17] In short, if wild mammals are an indication, bathing is unlikely to have been a big part of our ancient past.

As for our more immediate past, our human past, real bathing, in water, is both recent in our history and more varied among cultures and eras than might be superficially apparent. Bathing is one of those features of human culture that proves that history is not necessarily a story of progress, or at least progress as we tend to imagine it, progress as the steady change of societies of the past toward ways of life ever more like our own.[18] The Mesopotamians were not big bathers. Nor were the ancient Egyptians. The Indus River Valley people had a big central "great bath," but we can't be sure how they used it. It could have been for daily baths. It might have been for some sort of priestly, ritual ablutions.[19] But it also could have been the place they killed cows before eating them. Archaeology is tricky that way. The first Western culture that embraced bathing was Greek. Greek bath culture was expanded upon by the Romans. Superficially, it is this Greco-Roman pro-bathing culture in which we still find ourselves today, a culture in which bathing is viewed as not simply hygienic but also, in some real way, goodly, or even godly. We look at Roman baths and we are reminded of our own baths. We and the Romans are the same (except we substituted football for gladiator fights and they staged events in which naked emperors fought ostriches).[20] A clean life is a good life, a life to which the people of Western cultures have aspired since the classical period of Athens, a connection between our civilization and theirs. A well-bathed life is a good life. This is our subconscious mantra, one we wake up to each morning and embrace beneath the showerhead.

Yet, although the Greeks and Romans were both bathing cultures that valued hanging out naked in water, the water itself is likely to have been far from crystal clear. An excavation of the Roman baths at Caerleon, a site north of Newport in what is now Wales, discovered drains clogged with chicken bones, pig trotters, pork ribs, and mutton chops. These were the "light snacks" eaten "poolside." And though Romans generally regarded bathing as healthful and even recommended baths as treatments for some ailments, individuals with wounds were warned not to bathe because the dirtiness of the water might lead to disease.[21] The bathwater of Roman times would be more likely to cause disease than help prevent it.[22]

The Romans were far more likely to bathe, whatever the condition of the water, than those who followed them. The Visigoths, who came over the hill as the Western Roman Empire and Rome itself fell, with their shiny belt buckles and mustaches, were not much for baths. After the fall of Rome came a general shift toward less reading, less writing, and less infrastructure, including plumbing, and less bathing. This shift was persistent. It lasted, with local and largely ephemeral exceptions, from the end of the Western Roman Empire around 350 CE until well into the 1800s, which is to say nearly fifteen hundred years.[23] Not only did Europeans bathe very little during this period, many even forgot how to do so. The Romans made their own soap for bathing, but the daily know-how necessary to produce soap was forgotten in many regions, so little was the stuff used. In 1791, a French chemist named Nicholas LeBlanc invented a way to make soda ash (sodium bicarbonate) cheaply, which could be mixed with fat to produce a hard bar of soap. But even this more effective soap remained a luxury item. Bathing with soap or without was, at best, a monthly sort of affair and often not even that common. Nor was the lack of bathing strictly a tradition of the common folk. The kings and queens of Europe talked about their annual baths.[24]

The fall of the Western Roman Empire thus had many consequences, some of which lasted until long after the Renaissance.

With the Renaissance, art and science were reborn, but not bathing. Even Rembrandt's lovely mistress, dipping her ankles in the water, is unlikely to have done so very often, and she may not have stepped much deeper in, the preference for bathing being to wash the feet and the hands, but not necessarily the rest of the body. And, given that the water into which she was dipping was likely to be the same water into which chamber pots were emptied, the parts of her that weren't washed might have been more hygienic than those that were. Leave it to an ecologist to take the romance out of what seems to be an unambiguously romantic scene.

Overall, the question in the long history of bathing is why some people took it up again, rather than the reverse. Until very recently, most humans were unbathed. They would have smelled of the odors produced by the bacteria that grow on the human skin such as from armpit bacteria of the genus *Corynebacterium*. In the context of a city, the permanent funk rising up out of the pits of the people would have been matched only by the worse smells rising from other body parts. This would have been pungent in general, but especially when clothes weren't washed very frequently. We like to imagine, from our modern perspective, that given the opportunity, people would take a bath or stand beneath a watering can. But they didn't. Leeuwenhoek didn't. Rembrandt didn't. Then, in the 1800s, some people began to bathe regularly again. We can see the change as easily in the Netherlands, where it has been well studied, as we can see it anywhere. The answer has little to do with hygiene and much more to do with wealth and infrastructure.

In the early 1800s most water used in the cities of the Netherlands came from canals, rainwater collection, or, more rarely, wells. By that time, the surface water in urban and even many village canals was polluted with human and industrial waste. This pollution often affected shallow wells (much as it did later during the cholera outbreak in Soho in London) so much that the water smelled too bad to drink (as was also the case in London). Only the more affluent collected rainwater, and even then there was not typically enough water for daily use. Eventually, a few Dutch cities began the major shift to systems in which water was pumped from lakes and groundwater systems outside their boundaries into their centers. Two of the first cities to do so were Amsterdam, which had little of its own groundwater, and Rotterdam. In Amsterdam, it was necessary to pump water in to have enough for inhabitants and to stock ships traveling from the port. Rotterdam had enough of its own groundwater, but it had the problem that at low tide, the canals didn't have enough pressure to wash the feces out of the city. And so the city needed to pump water in, not so much as a source of drinking or other daily water but instead to flush the feces out into the sea.

Once water began to be piped into cities, it became a commodity. The rich accessed this commodity by paying for pipes to be laid directly to their property; the middle class, by paying for buckets of pumped water. In this context, it didn't take long for water and all that it could be used to do to become symbols of affluence. To be able to afford to flush away the odors of the toilet was prestigious. To be able to wash so often that the body did not smell was prestigious. The affluent installed water closets in their homes (to flush their waste) and then, more slowly, baths. Once this trend started, it never really stopped. It caught on in cities across Europe, cities in which to use a water closet was to be rich, to bathe was to be rich, and to be unable to bathe frequently was an honest indication of poverty or of the scarcity of clean water.[25] In time, showers were invented as a new means of "getting clean." In the coming years, this sense of cleanliness would be tied to the germ theory of disease and our desire to distance ourselves from all microbes in light of the knowledge that some cause disease. Since then, our desire to be clean and the amount of money we spend on getting clean have increased each year. Our desire to clean ourselves is fueled by a huge industry dedicated to convincing us we are dirty. We scrub, we buy sprays, and we stand, earnestly, beneath the shower's spray. Then we rub ourselves with salves. Billions and billions of dollars are spent not only to clean ourselves in ever newer ways, with ever more products, but also to make our bodies smell like flowers, fruits, or musk after we have done so.

What is rarely discussed is what makes our bodies or even water itself "clean." In the late 1800s in the Netherlands or London, *clean* meant that the water didn't smell, and when you used it, along with soap, on your body, neither did you. Once it was discovered that pathogens such as *Vibrio cholerae* caused disease, *clean* meant that water lacked these pathogens (or at least that such pathogens were rare). Later, *clean* would also come to mean free of dangerous concentrations of particular toxins. What *clean* has never meant, and will never mean, is sterile. Every bit of water that has ever sprayed down on you from the shower, risen around you in the bath, or poured into you from a glass or a sealed bottle has been full of life.[26] As is so often the case with the life in homes, what differs from one house or tap to the next is not the presence of life but instead its composition, which species are present and what those species do. This composition depends on where your water comes from in the first place.

THE STORY OF how water and the life in it get to our houses is both simple and extraordinarily complex. The simple part is the plumbing indoors. A water pipe comes into a house and branches into two. One branch travels to your water heater, where the water is warmed, before traveling again alongside the other pipe for water that is not warmed. The pair of branches then rebranch, in tandem, to get to each of your faucets and your showerheads.

The complexity has to do with what happens to the water before it reaches your house. The water's path depends very much upon where you live. In many parts of the world, water comes from a well sunk into the aquifer beneath a house, or it comes from a municipal water system that draws on an aquifer. *Aquifer* is a fancy word for the spaces in rocks that hold groundwater (where *groundwater* really just means water that is underground).[27] The groundwater in aquifers ultimately comes from rain. Rain falls on trees in forests, on grass in lawns, and on crops in fields. Over a period of hours, days, or even years (depending on local geology), rainwater seeps down into the soil, meter by meter. The infiltration of water into the Earth gets progressively slower the deeper the water travels. At great depths, the movement is very slow such that the water in the deepest aquifers might be hundreds or even thousands of years old. When you dig a deep well, you tap into ancient, untreated water. This untreated water then flows up and directly into a home. Or it goes to a water treatment plant. In many regions, such water treatment plants remove big material from the water (sticks, mud, and the like) and then send it, with little more in the way of treatment, to your house via underground pipes.

Water is safe to drink if it lacks pathogens (or has them in only very low concentrations) and if it has concentrations of toxins sufficiently low so as not to make us sick (with the concentration dependent on the toxin in question). The deeper and older an aquifer, the more likely the water is to be free of pathogens and, biologically, safe to drink. Much of the world's groundwater is safe to drink without any processing because of time, geology, and biodiversity. Geology influences the safety of the water inasmuch as some types of soils and rocks stop the spread of pathogens from surface waters. The biodiversity present in groundwater also helps to kill pathogens. Indeed, the more kinds of life present in groundwater, the less likely a pathogen is to survive. If a pathogen is a bacterium, it must compete for food, energy, and space. It must survive the antibiotics produced by other bacteria in the groundwater. It must avoid being eaten by predatory bacteria (such as species of the genus *Bdellovibrio*); it must also avoid being eaten by protists. Ciliates alone (such as which Leeuwenhoek encountered in his own water) can eat up to 8 percent of the bacteria around them a day. Choanoflagellates do even better and eat up to 50 percent of the bacteria around them a day. In addition, a pathogen must avoid being infected by bacteriophages, the specialized viruses that attack bacteria.[28] The organisms at the top of the food chain in these ecosystems tend to be small arthropods such as amphipods or isopods that have, like cave animals, lost their pigmentation and vision and move through the world guided by their senses of touch and smell. They include so-called living fossils that are relatively unchanged after millions of years of isolation and endemic species found nowhere else. These animals tend to be present only when groundwater is biodiverse, each species carrying out its function. Their presence is thought of as an indicator of water health.[29]

The life in groundwater ecosystems seems remote and obscure and tends to be studied only at a great distance (by scientists with long poles, drills, and nets). Yet it is estimated that 40 percent of the living mass, the biomass, of all bacterial life on Earth may be in groundwater. Forty percent! In some places, groundwater ecosystems are connected into a vast network of pockets, streams, and underground reservoirs. In others, they are disconnected underground islands. Just which life is present in a particular bit of groundwater is heavily contingent on where the water is, how old it is, and whether it is connected to or disconnected from other groundwater systems. Much as each oceanic island has its own unusual species, so too each groundwater system appears to have unique species found nowhere else. The deep waters of Nebraska and those of Iceland are different in part because the organisms living in the two aquifers from which that water comes have been evolving along separate trajectories for millions of years.

It may seem strange to drink groundwater that has not been treated with some biocide. But many of us do. Most well water is not treated with any biocide, nor is much of the municipal water of Denmark, Belgium, Austria, or Germany. The water of Vienna, for example, flows untreated directly from a karst aquifer. That of Munich is pumped from the porous aquifers of a nearby river valley directly through pipes and out of taps. The natural filtration of water by living organisms and time is thus of enormous benefit to humans. The trick is that it requires large spaces to be set aside where nature can do its work; it needs natural watersheds to be preserved. It requires time too. And it requires us to avoid polluting the groundwater with pathogens and toxins. Unfortunately, in

**Figure 5.3** An amphipod, *Niphargus bajuvaricus*, that lives in the groundwater in parts of Germany. This specimen was collected and photographed in Neuherberg, Germany. If this species tumbles out into your drinking glass, it is a sign that the aquifer out of which your tap water is flowing is probably healthy and biodiverse, a many-legged good omen. *(Günter Teichmann, Institude of Groundwater Ecology, Helmholtz Center Munich, Germany).*

many regions, we haven't set aside enough wild land for nature to do its work, or we have polluted groundwater, or, in some cases, there simply isn't enough groundwater available to supply large human populations. Under such circumstances, we must rely on human ingenuity to make water from reservoirs, rivers, or other sources safe to drink. Human ingenuity turns out to be a useful, but slightly crappy replacement for nature.

Human ingenuity relies heavily on biocides. Beginning in the 1900s, water treatment plants in some regions began to use chlorine or chloramine to kill bacteria in water, with the aim of controlling pathogens. This was necessary in regions where the aquifers had become polluted. It was also necessary in the many regions where aquifers were insufficient to provide for growing human populations and water needed to be piped not from the ancient depths but instead from shallow rivers (such as the Thames in London), lakes, and reservoirs. In the United States, all municipal (city) water is now treated with biocides at treatment plants.[30] In addition, because the pipes in water systems in the United States tend to be old relative to those in continental Europe and elsewhere, they leak and water stagnates.[31] Whereas in natural aquifers older water is better water, the opposite is true in our pipes. In pipes, stagnation can favor the growth of pathogens. To counter such stagnation, water in the United States is typically treated with extra biocide as it is leaving the treatment plant beyond that which is used in similar European treatment systems. Sometimes the biocide is chlorine, sometimes chloramine. Sometimes it is a mix of these two. Water treatment plants can be technologically sophisticated and yet, to the extent to which nearly all rely primarily on getting rid of life through a series of sieving steps (through sand, through carbon, through a membrane), sometimes a dose of ozone, and then killing microbes with a biocide, they are really simple.[32] Meanwhile, even after disinfection with biocides, the water leaving treatment plants is not sterile. Instead, it is water in which the most susceptible species have been killed and the toughest species have survived, alongside the dead bodies of the susceptible species and the food those susceptible species were eating.

If ecologists have learned anything in the last hundred years, it is that when you kill species but leave the resources upon which they feed, the tough species not only survive but thrive in the vacuum created by the death of their competition. They enjoy what ecologists call "competitive release." They are released from competition, released too, often, from parasitism and predation. In the case of water systems, we would predict the species that thrive to be those that are resistant to or even just slightly more tolerant of chlorine or chloramine. Mycobacterial species tend to be very tolerant of chlorine and chloramine.

As NOAH AND I, along with our other collaborators, began to consider the data from the showerhead study, we kept the differences between untreated groundwater, treated municipal water in the United States, and treated municipal water in Europe in mind. Medical researchers have predicted that mycobacteria might be more common in well water inasmuch as it is less controlled, less treated, more susceptible to nature's whimsy. But as ecologists, Noah and I, along with the rest of our team, also had to contemplate the opposite—namely, that mycobacteria might actually be more common in the showerheads of people with municipal water, particularly that from treatment plants and countries that use chlorine or chloramine, particularly water from such plants in the United States. Mycobacteria are relatively resistant to chlorine and chloramine. Perhaps tap water was treated with enough biocide to kill most species, most species except the mycobacteria. We found some precedent for this idea. One study of bacteria in showerheads had already noted that cleaning an individual showerhead from a Denver shower with bleach solution led to a threefold increase in the abundance of a species of *Mycobacterium*.[33] It was an anecdote, but an interesting one.

When we looked at our data, we expected only about half a dozen different kinds of mycobacteria from showerhead samples. We expected to find those species cultured in medical study after medical study. Instead, we found dozens of species, quite a few of them apparently new to science. Just which species were present in a showerhead appeared to depend in part on region. Different species dominated in Europe than in North America (and not just because of the different types of showerheads). But even within the United States the species present in Michigan were different from those in Ohio, which were different again from those in Florida and Hawaii. These differences might be due to the different aquifers from which the water comes, or whether the water comes from an aquifer or surface water, or even some aspect of climate or ancient geology.

Yet, while the identity of the species of *Mycobacterium* present in a showerhead was hard to account for, the abundance of mycobacteria in the showerheads was more predictable. We measured the amount of chlorine in the tap water of the house of each of our participants. The concentration of chlorine in the tap water from homes using municipal water in the United States was fifteen times higher than that of homes with well water. This was enough of a difference, we thought, to have an effect. But we expected a modest effect. The effect was huge. In the United States, mycobacteria were twice as common in municipal water than in well water. In some showerheads from municipal water systems, 90 percent of the bacteria were one or another species of *Mycobacterium*. In contrast, many of the showerheads from houses with well water had no *Mycobacterium*. In place of *Mycobacterium*, the biofilms from houses with well water tended to have a high biodiversity of other kinds of bacteria. In Europe, the abundance of mycobacteria in showerheads from well water systems was low, just as in the United States. But the abundance of mycobacteria in Europe was also low in showerheads from houses with municipal water (half that of municipal systems in the United States), as might be expected given that many European municipal water systems do not use biocides at all. In our samples the residual chlorine measures in European tap water were eleven times less than in tap water in the United States. As we were pondering these results, Caitlin Proctor at the Swiss Federal Institute of Aquatic Science and Technology in Switzerland published a new study very much in line with what we were finding. Proctor and her colleagues compared the biofilms of the hoses that lead into shower heads from seventy-six homes around the world. They found that samples from cities that did not disinfect their water (including samples from Denmark, Germany, South Africa, Spain, and Switzerland) tended to be thicker (more gunk), but samples from those that did disinfect their water (including Latvia, Portugal, Serbia, the United Kingdom, and the United States) were more likely to be lower in diversity and more dominated by mycobacteria.

So far, our results match Caitlin Proctor's results, which match what we expect if the subset of water treatment plants that use biocides kill many species and, in doing so, create conditions that allow mycobacteria to thrive. If true, this would mean that our fanciest water treatment technology is creating water systems filled with microbes that are less healthful for humans than those found in untreated aquifers (or at least that subset of untreated aquifers that has been deemed safe). We couldn't explain all of the variation in the abundance of mycobacteria among houses. Nonetheless, we hypothesize that, in general, chlorine and chloramine use increases the abundance of mycobacteria in showerheads, which makes it more likely that people will develop a mycobacterial infection. In our analysis, the mean abundance of the most pathogenic strains and species of *Mycobacterium* in showerheads in a particular state was highly predictive of the number of mycobacterial infections in that same state, predictive of the pattern shown in Figure 5.1. But there are twists in the story already. One of the twists is Christopher Lowry.

Lowry has spent twenty years studying one particular *Mycobacterium* species, *Mycobacterium vaccae*. He and his colleagues have found that exposure to this *Mycobacterium* species boosts production of the neurotransmitter serotonin in the brains of mice and humans. Increased serotonin production tends to be linked to greater happiness and reductions in stress. Indeed, Lowry has shown that, at least in mice, inoculating individuals with *Mycobacterium vaccae* leads them to be more resilient to stress. Working with a colleague, Stefan Reber, in Germany, Lowry tested this by inoculating average-size male mice with *Mycobacterium vaccae*. He then introduced those mice, as well as average-size male mice without the bacteria (the control mice), into a cage with aggressive, sumo-size male mice. Afterward, he tested the stress-related compounds in the blood of the average-size male mice. The control mice pissed themselves, cried softly into their wood shavings, and registered high on every stress test. The males that had been treated with *Mycobacterium vaccae* weren't stressed at all. Conversations are now ongoing about whether soldiers could be inoculated with *Mycobacterium vaccae* before they go to war to reduce the risk that they might suffer from posttraumatic stress disorder (inasmuch as they almost certainly will be exposed to traumatic stress). All of this sounds a bit crazy, but even in these early days, it has been recognized by Lowry's peers as very important work. In 2016, the Brain and Behavior Research Foundation, for instance, ranked the work as one of the top ten (out of five hundred) contributions by researchers that it funded.[34] Lowry suspects that many *Mycobacterium* species may have effects similar to those he has observed of *Mycobacterium vaccae*. The only way to know for sure is to test them one by one, and so this is what Lowry is now doing. He is culturing the mycobacteria we have gathered in showerheads to see whether any other species behave like *Mycobacterium vaccae*. If they do, it may mean that some of the *Mycobacterium* falling on you from your showerhead may be beneficial in reducing your stress.

The showerhead is one of the simplest ecosystems in your house. The average showerhead has dozens, and at most hundreds, of species in it rather than thousands. Even so, Lowry's research reminds us that sorting out just which kinds of microbes are good and which are bad is gnarly, convoluted, and hard. Some mycobacterial strains may make you sick; others may make you happy. Until we sort out which is which with some confidence, our results will prove totally dissatisfying to our participants (perhaps to you too). They are also dissatisfying to us. That is the thing about science. One imagines we do it out of joy and curiosity, and that is part of it, but sometimes we do it out of frustration. Sometimes, we do it because it is just so incredibly frustrating to not know an answer, even about something so immediate as our showerheads, that we have to go back into the lab and keep working, because the idea that no one yet knows what is going on, well, it keeps us up at night.

But what, then, should you do about your showerhead? We don't know, but I'll tell you what I think. Check back in a year or so and see whether I was right. I think that while some *Mycobacterium* species can be beneficial, the average species is at least a little bit of trouble, particularly for immunocompromised individuals. I think that these bad-news *Mycobacterium* species become more common the more we try to kill everything in our water and, in doing so, kill off *Mycobacterium*'s competition. We have shown that plastic showerheads tend to have less *Mycobacterium* than do metal showerheads, as might be expected if other bacteria are able to metabolize the plastic and, in doing so, outcompete the mycobacteria (Caitlin Proctor found a similar pattern in the hoses of showerheads). Finally, I think that the water that is healthiest for bathing is that which comes from aquifers rich with underground biodiversity including crustaceans. The crustaceans in these aquifers are an indication not of the dirtiness of the water but of its health. The trick is that these aquifers require time, space, and biodiversity to work. Also, they have to remain free of pollution. I suspect that this is a hard lesson for big cities to take to heart and so, over the coming years, I think we will try to kill everything in our water systems. Unfortunately, in doing so, we'll also accidentally favor tough species (such as *Mycobacterium* and *Legionella*) that we don't really want to have pouring over us quite so much. Meanwhile, we will begin to study natural aquifers in more detail and find that they differ in how effective they are at preventing toxins and pathogens from building up in our water systems. Having figured that out, we'll try to replicate those natural aquifers. We won't be great at it, but we'll slowly figure out how to do a much better job than we are doing today, and the key will prove to be (as it often is) valuing biodiversity, valuing the work that nature does oh-so-much-more-effectively than we do. As for whether it is worth buying a new showerhead every so often, we don't know yet. But I suspect that after reading this you will go home and change your showerhead anyway.

# 6

# THE PROBLEM WITH ABUNDANCE

> What would an ocean be without a monster lurking in the dark?
>
> —WERNER HERZOG

GENERALLY SPEAKING, we tend to dislike species that succeed, unless of course we can eat them. We now control so much of the planet that species that succeed almost invariably do so at our expense. They eat us or our food or the things we have made, such as our homes. Since we first built homes, species have been nibbling them back to the ground. In the story of the three little pigs, it was the wolf that knocked the houses down because he was after the pigs. In real life the species that knock our homes down are far more likely to be far smaller than wolves and yet no less dangerous. Just which species threaten our homes depends on how and where the house is built. Stone houses can last thousands of years, which is why some of the buildings from the very earliest civilizations are still standing. Mud houses, too, can last, so long as conditions are dry. But most of our houses are made of dead trees, and many species can eat dead trees. Termites, of course, can bite into wood;

they then rely on specialized bacteria in their guts to digest the wood. But the grand masters of destruction are fungi.

In a dry house, fungi can be relatively inconspicuous. But when water pours onto walls or floors, it allows fungi to grow. Fungi creep up the gradient of moisture, eating. If you could hear them, the sound would be terrible, the sound of their hyphae drilling holes in the cells of ancient wood, cracking them open one by one. Fungi eat and creep using their hyphae. As fungi contract hyphae in one region and expand in another, they can actually move from place to place; they crawl in slow motion. For fungi, your walls are full of nutrients. Fungi can eat nearly everything a wooden house is built of so long as they have enough water and time. Fungi eat wood. Fungi eat thatch (fungi also compete with bacteria for the little bits of food present in dust). Given hundreds of years, fungi can even release chemicals that break down bricks and stones. In their growth, everything fungi do becomes magnified. They more quickly degrade wood and paper. They produce more spores, more toxins, more everything. When abundant, some fungi can eventually turn a house back to soil, much as they would a log. But long before that happens, they can also cause other problems. They can be dangerous if accidentally consumed. Some fungi can trigger allergies and asthma. Then there is *Stachybotrys chartarum:* toxic black mold. Toxic black mold can reach great abundances in houses. When it does, it seems, often, to be at our expense.

If we understand any fungus species in homes, this conspicuous mold should certainly be one of them. *Stachybotrys chartarum* should not be a species that offers us surprises. If you see *Stachybotrys chartarum* in your house, most housing professionals will advise you to call a mold abatement company. Such companies will come in and get rid of all of the visible *Stachybotrys chartarum* fungus in your house. Your books may be scrubbed and scrubbed (or even thrown away), your clothes treated or maybe thrown out too. This drama is reenacted again and again. The details and protagonists differ. The perceived villain stays the same, but so too does a barbarous ambiguity about what is actually going on.

Despite having spent years reading about and thinking about fungi, I didn't really understand the story of *Stachybotrys chartarum* until I met Birgitte Andersen. Birgitte is an expert on the fungi of houses. She studies two things: what eats the building materials in houses and how those species, species most would regard as sinister but that she considers fascinating, get into houses in the first place. Birgitte spends a lot of time with *Stachybotrys chartarum*.

I emailed Birgitte asking to meet, and she invited me to travel from central Copenhagen, where I was staying, to her university, the Technical University of Denmark (DTU). I traveled by bike. The day was, relatively speaking, a sunny Danish day, which is to say that when I parked my bike at her building, I was soaked from the rain. In my damp clothing, I felt fungal. We would talk about fungi. The ambiance was perfect, albeit discomfiting.

Birgitte's office is on the second floor of a building dedicated to technical science, the use of fancy equipment to solve applied problems. In this building, Birgitte is a hold-out. She *loves* fungi. She is dedicated to the study of fungi. For work, she grows fungi and then carefully, painstakingly, identifies them under a microscope, takes pictures of them, and adds them to her guide of the common and rare fungi of Denmark. After work, as a hobby, she does much the same, just not for pay. She finds the fungi beautiful and each fungus beautiful in a different way. Fewer and fewer people each year seem to have the skills or the obsession necessary to grow and identify fungi; she has both. Once, she had many colleagues who shared her passion, people to whom she could run around the corner to see and say, "You won't *believe* this fungus I am seeing." But Birgitte's colleagues with a passion for fungi have retired. And at Birgitte's university, like many others, few new biologists with the ability to actually grow, identify, and catalog organisms—in this case, fungi—are being hired. One article in the magazine *The Scientist* went so far as to ask whether scientists with expertise in naming, classifying, and growing wild species were going extinct (the conclusion was yes).[1] Such work is essential. The vast majority of fungal species are not yet named. But the effort to catalog species and their basic biology lacks glamour and so is less likely to be rewarded by hiring committees and funding agencies. Birgitte now stands alone, at the end of her hall, the last person in her building able to identify fungi well, and one of the last few in Denmark.

By the time I went to visit Birgitte, Noah Fierer and I, along with our collaborators, had worked with the public to sample dust from the door sills of more than a thousand houses. From that dust, we identified the species of bacteria in each sample by decoding their DNA. Later we also did the same for fungi and found a ferocious diversity of fungi in homes and on homes. We found forty thousand kinds of fungi.[2] Numerically, this was fewer kinds than for the bacteria and yet it was a bigger surprise. Fewer than twenty-five thousand species of fungi—mushrooms, puffballs, and molds—are named in North America. We found a greater diversity of kinds of fungi (or at least kinds of fungal DNA) in homes than have yet been named from North America, indoors or out. Thousands of the species of fungi we discovered in homes are likely not yet named. These nameless fungi spoke to our ignorance not only of what is in homes but also more generally. As for the named fungi, each had a unique story. Inasmuch as the life cycles of fungi are often dependent on other species, the fungi indicated not only their own presence but also the presence of the organisms on which they depend. Some of the fungi were pathogens of grapes and suggested the presence of vineyards. Others were pathogens of specific species of bees (and implied the presence of those bees). Others were parasites able to take over the brains of some (but not other) ants.[3] In eastern North Carolina, we found fungi of the genus *Tuber*, which form symbioses with the roots of trees and then, in order to disperse from one place to another, produce truffles that mimic the pheromone produced by male pigs to attract female pigs. Wooed by the truffle, the female pigs dig it up, eat it, and then, if the truffle is lucky, poop it somewhere else in the woods near a young tree not yet colonized by truffles.

For bacteria in our homes, the emerging story is one in which we have sealed out most environmental bacteria (to our detriment) and instead have surrounded ourselves with bacteria able to deal with extremes, such as the conditions in showerheads, or with our food or our bodily waste. Superficially, because both fungi and bacteria tend to be lumped together as "microbes" along with many other small life-forms, one might assume the same for fungi. But fungi are actually much more closely related to animals than they are to bacteria, so much more closely, in fact, that one of the challenges of controlling fungi is that chemicals that kill fungal cells tend to also kill human cells. Also, unlike the case for bacteria, very few fungal species live on human bodies, whether as pathogens or mutualists. Our bodies are too warm for fungi (it has been argued that warm-bloodedness itself evolved as a way to keep fungi at bay).[4] The story of fungi in homes, then, could be entirely different from that of bacteria. It has proven to be.

Many of the fungi in houses appear to be species that have simply drifted in from the outdoors. The fungi in houses are very similar to those we find on the outside of houses. Different fungi are found in houses in different regions primarily because the fungi outside of the houses are different.[5] The effect of the outside fungi on the fungi inside is so great that we can identify the origin of a dust swab in the United States within fifty to one hundred kilometers based solely on which fungal species are on the swab.[6] Swab your house, send us the swab, and we can tell you where you live (though if you do this, also send a couple of hundred bucks; this turns out to be an expensive party trick). With regard to these many thousands of species, then, the best way to change your exposure to them—really, probably the only way—is to move.

In addition to species that drift in, we also found species that seemingly specialized in life in homes, species more common indoors than outdoors. But we found so many of these species it was hard to know which ones to focus on, hard to know which were best able to move with us from place to place and thrive in our presence. To gather some more insights, I turned once again to the International

Space Station (ISS), along with the Russian space station, the Mir. We know that any fungi found on the space stations are really living there, indoors. They can't have floated in through an open window or hatch, as the case might be. Not even fungi can survive very long in the conditions outside of the space stations.[7]

We know the most about the fungal life on the Mir. Ever since its first launch in 1986, Mir was sampled again and again. Five hundred air samples have been taken to check for fungi. Another six hundred samples were taken from surfaces around the space station. These samples were then cultured, either on Mir itself or back on Earth. The samples weren't cultured exhaustively,[8] and yet even so, the results were unambiguous: Mir was a fungal jungle. It was full of more than one hundred different species of fungi. Fungi were found in all but a handful of the more than a thousand samples taken from Mir.[9] These fungi were alive and metabolizing, too, so much so that one cosmonaut described Mir as smelling like rotten apples (which is perhaps better than the body odor smell of the ISS). As if that weren't bad enough, at one point the Mir lost contact with Earth. A communication device broke down; the insulation around its wires, it was later revealed, had been eaten by fungi, so the wires shorted out.[10] The fungi have been far more successful, in other words, in establishing themselves in space, having sex and living out many generations, than humans have been. Here, then, is a cautionary tale for any potential attempt to colonize Mars. Long before humans successfully colonize, live on, and have children on Mars, fungi will have done so.

Initially, the ISS was described as being, relative to Mir, if not sterile at least less fungal. Sure, the Mir was colonized by fungi, but it had a reputation for being held together by duct tape and dreams, so maybe that wasn't surprising. But with time, the life on the ISS also grew diverse and fungal. By 2004, thirty-eight species of fungi were found to be common on the ISS. These thirty-eight species were largely a subset of those found earlier on Mir, which were in turn a subset of those we find in homes.

Many of the fungi found on spacecrafts are described as "technophiles" by the biologists who study them because of their ability to degrade the metals and plastics out of which space shuttles and space stations are made.[11] "Technophiles" sounds to me like the name of a boy band playing synthesizers, but it is meant to denote that these species like ("phile") technology; they like it so much that they eat it.[12] The species already shown to be feeding on the ISS itself include *Penicillium glandicola* (a relative of bread mold), species of *Aspergillus* (relatives of the organism used to make the Japanese rice wine, sake), and a species of *Cladosporium*. Not all of the fungi on board are technophiles, though. On Mir, but not on the ISS, brewer's yeast (*Saccharomyces cerevisiae*) was found (perhaps suggesting that the Russians had a better time out in space).[13] The researchers also found species of the genus *Rhodotorula*, the pink fungus often seen growing in grout, on shower walls, and, ever so rarely, in toothbrushes and on humans on Earth.[14] Here, then, living among the astronauts, were species that really, definitely thrive in the conditions indoors.[15]

We found all of the kinds of fungi present on the space stations in homes. In fact, the species of fungi that were present in the space stations were present in virtually every home we sampled. Just which species were most common depended on the house. Houses with more people in them tended to favor fungi associated with human bodies or foods.[16] The way in which a house was heated or cooled also influenced which species were present. In particular, houses with air-conditioning tended to be more likely to have *Cladosporium* and *Penicillium* fungi. These fungi (to which some people have allergies) grow in the air-conditioning units themselves and then spread through homes and offices when the air-conditioning is turned on.[17] When you turn your air conditioner on in your house or car and smell an unusual odor, it is the odor of these fungi exhaling.[18]

We will be disentangling the mysteries of our data on household fungi for decades, but one mystery demanded more rapid consideration, the mystery of a species that was absent inside the space stations and rare in our samples of homes, *Stachybotrys chartarum*. *Stachybotrys chartarum* is conspicuous as a problem, and yet inconspicuous in our samples. The absence of *Stachybotrys chartarum* on the space stations might result from the absence of its food source. The ISS, to my knowledge, is devoid of wood or even cellulose, though one might expect it to be able to degrade some of the plastics.[19] But this doesn't explain its rarity in our study of houses.[20]

I asked Birgitte about this mystery. I explained our study. I didn't specifically mention the ISS, but I was thinking about it, floating above us as we talked, distant and yet nonetheless fungal. Birgitte was unsurprised. "Its spores are heavy and held out on sticky slime heads. Why would you find it?" In other words: if it doesn't float in dust, it shouldn't be in dust. And then, for emphasis, "Why were you even thinking you might find it?" Birgitte is direct. Why would we indeed. But how, then, I asked her, was it getting into houses if it wasn't floating in? How was it getting into houses and why had it failed to get into the space stations (when so many other species seem to have no trouble)? "We have," she said, "done a study that might interest you."

As we snacked on cookies and nuts foraged from a drawer (each invisibly and inadvertently powdered with a diversity of fungi from the air we were both breathing), Birgitte told me about her study. It focused on the materials out of which modern homes are made: drywall, wallpaper, wood, and cement. Birgitte is not terribly interested in the air in houses. She is interested, instead, in the materials, the pieces out of which houses are built: their bricks, their stones, their sticks, and, especially, their drywall.

Each building material in homes, Birgitte found, seemed to have its own kinds of fungi—much as might turn out to be true on the space stations, too, if their materials were studied in enough detail. On the cement, Birgitte found fungi of the same sorts one might find on the ground outdoors, a slurry of soil life, including some of the very first species of fungi ever to be studied by scientists.[21] These fungi were studied by scientists because they were at hand; they were at hand because they lived in the scientists' homes. She found *Mucor*, for instance, which Robert Hooke depicted in *Micrographia*, the book likely to have inspired Leeuwenhoek. She found *Penicillium*, which Alexander Fleming found by chance in his laboratory (just another building, after all) where he discovered antibiotics. *Penicillium* uses these antibiotics to weaken the cell walls of the bacteria with which it competes for food, causing the bacteria to explode as they try to grow. We use the same antibiotics to fend off bacteria pathogens such as *Mycobacterium tuberculosis*, with which we fight for our own survival.

These fungi, *Mucor*, *Penicillium*, and the like, were also kinds of fungi that had made it onto a space station.[22] That they live both on cement floors and in space stations means that they are fungi we probably need to find a way to live happily alongside. They got past NASA's control measures, and if they could hail a ride to outer space, they can presumably hail a ride nearly anywhere else too.[23] They may well be the same species that grew on the walls of our ancestral caves; if so, they have traveled from those caves with us wherever we have gone. These are some of the same fungi that, given enough time, eat away at bricks and even stones. On the floors of houses, they may be eating, too, slowly, or they may be using the cement as a habitat (their hyphal fingers holding on) while they actually eat bits of dirt too small to notice or glues and other materials on the cement's surface.[24] These fungi pose problems for people who want to preserve monuments for hundreds of years, but in your basement they are likely to be little more than interesting evidence of the power of fungi to devour, if given time, nearly anything.

On the wood, too, there were fungi. We build many of our homes out of wood and long have. But wood is biodegradable. It is composed of both cellulose and lignin. Cellulose is the stuff of paper; lignin the sturdy stuff that keeps the roof up. Many microbes can break down cellulose, but only fungi and a handful of bacteria are able to break down lignin.[25] The fungi Birgitte found on wood in our homes included species that make the enzymes for breaking down at least cellulose and, in some cases, lignin.[26] The surprise is not that this also occurs among our two-by-fours and beams but instead how long we are able to prevent it from occurring. Many of the species of wood-degrading fungi that live in homes simply blow in from outdoors, such that their precise composition is determined by the kinds of trees out of which these homes are made as well as by the sorts of forests nearby. Other species, such as the dry rot fungus *Serpula lacrymans*, are known to have been carried on ships with humans around the world.[27] They followed us as we built, again and again, homes made out of their food. They follow gratefully.

It was when Birgitte considered drywall, wallpaper, and plaster covered with paper (and then painted) that things got even more interesting. These substrates were, when wet, full of fungi.[28] What was more, those fungi included the toxic black mold *Stachybotrys chartarum* a full 25 percent of the time. This is even an underestimate of the proportion of wet houses in which *Stachybotrys chartarum* might occur. After all, Birgitte took just small samples from each home. *Stachybotrys chartarum* isn't rare in wet drywall, then. It is so common as to be the ordinary, expected fungus to occur when drywall gets wet. The mix of water and cellulose available in drywall and wallpaper appears to be a perfect substrate for *Stachybotrys chartarum*. This was a discovery, a big discovery, but Birgitte still needed to explain how the *Stachybotrys chartarum* was getting into the drywall in the first place.

*Stachybotrys chartarum* does not float through the air. As far as anyone knows, it does not ride on or in termites or other household insects either. In theory, it might be brought indoors on clothing. Rachel Adams, an expert on indoor fungi at the University of California, Berkeley, learned firsthand about just how many species can ride on our clothes. In one of the most careful studies of the fungi in buildings to date, Rachel found that one of the fungi she detected in a conference room of a university building was brought there inadvertently by a lab mate who had recently visited a mushroom event, where she handled puffballs.[29] Fungi ride lab mates. But Birgitte wasn't really concerned about clothing. She was thinking about building supplies.

What if the mold was in the drywall all along? What if it was being introduced into the drywall when the drywall was being made and then sat there happily, quiescent, until the drywall got wet? This is just what Birgitte proceeded to test, this radical idea, this idea that, if correct, could put her at odds with the multibillion-dollar drywall industry. As she started to look, she realized she wasn't the first one with this idea. An earlier paper had also suggested the same possibility, but the earlier paper had not tested it.[30] She would.

In the United States, academics have some degree of freedom in research, but increasingly it seems to be less than absolute, in no small part because of the great power of corporations. This is not to say academics don't publish dangerous ideas, ideas hazardous to governments or businesses. It is to say that many American academics have seen enough Hollywood movies to ponder the possibility that one ought to carefully consider the consequences of research at odds with the economic incentives of powerful business leaders.[31] It is possible that the same worry creeps into the minds of many of Birgitte's Danish colleagues when they do challenging work. But when I asked her about this risk, Birgitte voiced little concern for the possibility that there might be negative consequences to studying what lives in drywall produced by companies that have an enormous stake in maintaining the status quo (the status quo for drywall, anyway). She just wanted to know what, if anything, was in there. Her emotions were uncomplicated. She was curious. So she looked.

First, Birgitte studied pieces of brand-new drywall, thirteen sheets in total, from four different hardware stores in Denmark. She chose two brands of drywall among those thirteen sheets and three different types of each brand (fire resistant, moisture resistant, and regular). She then cut multiple circular discs from each of the sheets and dipped the discs in ethanol (or, in alternative protocols, just to be sure, bleach or Rodalon), which sterilized their surfaces. Then, for seventy days, she soaked the surface-sterilized samples in sterile water so that any fungi inside the samples might grow. It seemed unlikely anything would really be alive in the dry, brand-new drywall. It was a painstaking longshot—one dependent on careful, tedious work, including the simple but daily task of checking each and every disc for fungi.

Finally, one day, she saw growth. Then, more growth. Birgitte found lurking inside of brand-new drywall the fungus called *Neosartorya hiratsukae*. This fungus has recently been implicated in the complex mix of causes of Parkinson's disease. It is unlikely to be the sole cause of the disease, and yet its presence is nonetheless not good news. *Neosartorya hiratsukae* was on every single sheet of drywall, regardless of type, regardless of which store it came from, and regardless of which company it was made by. Birgitte also found the fungus *Chaetomium globosum*, an allergen and opportunistic pathogen.[32] It was on 85 percent of the pieces of drywall. And then there it was, black and potent, *Stachybotrys chartarum* on half of the samples.[33] Once it started to grow, it covered the drywall discs, darkening them with life. Nor were these the only species present. Eight other kinds of fungi were also found inside the drywall, waiting.

Now came the real test of whether Birgitte harbored any more fear of the drywall companies than she admitted. Would she publish the results—results that implied the drywall industry had some role in influencing which fungi arrived in your house, some role in potentially affecting your health? *Stachybotrys chartarum* is often linked to health problems. *Neosartorya hiratsukae* can be a human pathogen. It is rarely detected on wet drywall in homes but is also hard to spot. It produces small, white fruiting bodies the color of the drywall itself. Because the fungus showed up in each sample, regardless of the store it came from, Birgitte's results clearly implicated the drywall manufacturing companies. Of course she would publish the work. "What would they do?" she asked me. "Take my job? Then who would identify the fungi?" So it is that we now know, without a doubt, that the fungi in the drywall in homes comes preloaded in brand-new drywall. Birgitte is now working to find ways to kill these fungi in drywall before it is sent to new homes. There is unlikely to be any easy way to kill it in drywall that has already been installed. Any treatment that would kill the fungus in drywall hung in houses is likely also to destroy the drywall and be hazardous to people. Meanwhile, the fungi wait for moisture. Their patience is great.

It is unclear just how the fungi are getting into the drywall, but it is possible that when recycled cardboard is stored for use in drywall production, it becomes a hotspot for the growth of fungi. When the cardboard is then ground up and incorporated into the drywall, the fungi survive the process as spores. Perhaps, Birgitte imagines, the cardboard could be treated in some way. But it isn't yet. And so, if Birgitte is right, the drywall that comes to your home today is still arriving with the fungi already present. It's fine; as Birgitte notes, just don't let drywall get wet.

Knowing how *Stachybotrys chartarum* and other heavily spored fungi get into houses isn't all we need to understand to make sense of the fungi in houses. Even though Birgitte has identified, it appears, how these particular fungi are entering homes, she hasn't really identified where these fungi evolved, their native region and natural habitat. The closest relatives of *Stachybotrys* seem to be species of *Myrothecium*, from the tropics, but we know almost nothing about *Myrothecium*, including whether it occurs in houses in the tropics. It is speculated that many species of relatives of *Myrothecium* and *Stachybotrys* exist unnamed. In rural environments, *Stachybotrys chartarum* has been found in piles of grass, but this probably says more about where we have looked than it does about the biology of *Stachybotrys chartarum*. Soil, it is said, may also be the native home of *Stachybotrys chartarum*, but this too is so vague a prediction as to lack meaning. Then, there is the question of what disperses *Stachybotrys chartarum* in the wild, what carries it from place to place. Beetles or ants might, but these are guesses. No study has ever tested whether these or any other insects carry

*Stachybotrys chartarum* spores. Nor do we know how long *Stachybotrys chartarum* has been associated with homes (it would be great to know which fungi are found in traditional homes around the world, or in the bits of houses found at archaeological sites, but again, no studies have yet been done). And then we are also left with the question of just how dangerous to us the fungi in homes might be. After all, we spend billions of dollars remediating these fungi. We tear down houses. Formerly healthy people live lives of desperation trying to cure diseases they are told are due to *Stachybotrys chartarum* fungi exposure, largely to little avail. It is still very hard to say.

For good reasons, no one has done the experiment in which one inoculates a house with *Stachybotrys chartarum* and then looks at the effects on the family therein. Nor has anyone taken houses and wet them to see if (or when) the *Stachybotrys chartarum* grows and illnesses begin. There are two ways that this fungus might make us sick, though. It could poison us with its toxins, or it could trigger and exacerbate allergies and asthma.

First, the toxins. We know that *Stachybotrys chartarum*, like many fungi, produces scary compounds called macrocyclic trichothecenes and atranones. *Stachybotrys chartarum* can also produce hemolytic proteins. If eaten by sheep, horses, or rabbits, these compounds, the proteins especially, cause leukopenia (a deficiency in white blood cells). It has been speculated that these same proteins could also cause pulmonary hemorrhage in human infants. Mice whose noses are injected with the spores of *Stachybotrys* suffer, though just how much they suffer depends on which strain of *Stachybotrys* they get. Mice given a strain that produces more toxins suffered "severe intra-alveolar, bronchiolar and interstitial inflammation with hemorrhagic exudative processes." Put more plainly, their lungs became inflamed and started to bleed.[34]

But just because *Stachybotrys* can produce toxins doesn't mean it necessarily does so in houses. Recently, Birgitte and colleagues developed a new method of detecting the presence of toxins from

*Stachybotrys chartarum* in dust. In doing so, they showed that the more *Stachybotrys chartarum* was present in rooms of a kindergarten in Denmark, the more of its toxins were present in the dust in those rooms. It is not yet known if the same is true more generally. It might be.[35] But to develop an illness would require eating (or, like the mice, sniffing) large quantities of the fungus. An infant in a home in which *Stachybotrys chartarum* was present, growing abundantly, *and* producing toxins might incidentally consume large quantities of the fungus and get sick in the way that lab mice and domesticated animals get sick. So far, such a case has never been documented. *Neosartorya hiratsukae* may be more likely to cause serious health effects from toxins than *Stachybotrys chartarum*, but it is even less well studied (it is no less common, but much less conspicuous). All of these complexities make Birgitte, one of the world leaders in expertise on *Stachybotrys* and its consequences, say that she hates when people ask what we know about the health consequences of the toxins produced by indoor fungi. As she puts it, "It is just so damned complicated and difficult to prove."

But even if *Stachybotrys* toxins only rarely make people sick, the fungus could potentially still affect us in other ways. When inhaled, *Stachybotrys chartarum* can trigger allergies. The blood of a relatively high proportion of people shows evidence of an allergic response to *Stachybotrys chartarum*. In these cases, the exposure to *Stachybotrys chartarum* may have been outdoors, but in others it is probably from *Stachybotrys chartarum* grown on drywall in wet houses. In this, *Stachybotrys chartarum* is not alone. Many other fungi, including fungi that become more common when houses are wet, trigger allergies and asthma.[36] The authors of the biodiversity hypothesis, Hanski, Haahtela, and von Hertzen, would argue that what is happening is probably that our lack of exposure to diverse environmental bacteria is making our immune systems more likely to develop allergies. I think they are probably right. If they are, in houses where fungi or other organisms (such as German cockroaches or dust mites) are abundant, those abundant organisms then serve as triggers. The trigger doesn't matter, I hypothesize, except where a lack of exposure to a diversity of bacteria, or some other precondition, sets the stage.

If the biodiversity hypothesis is right, we might expect the correlations between the presence of abundant fungi in a home and allergies to be complex and contingent. Indeed, whereas some studies find that people in houses with more fungi or more allergenic fungi are more likely to suffer from allergies or asthma, a far greater number of studies show no effect.[37] But it might be that reducing asthma and allergy symptoms once they appear is simpler than understanding why and when these diseases emerge in the first place. That seems to be the case. A team, led by Carolyn Kercsmar at Case Western Reserve University, found 62 children who had symptomatic asthma and were living in houses with indoor mold. Kercsmar then randomly assigned the children and their families to one of two treatments. The families of half of the children (the control group) received instructions on how best to manage the asthma, and nothing more. The families of the other children (the remediation group) received the same instruction and the study team went into their houses and removed wet wood and drywall, replaced it with new, dry material, stopped the flow of water into the home, and made alterations to air-conditioning units. After the intervention, the concentration of fungi in the air of remediation group houses decreased by half. The concentration in the control houses was unchanged. More significantly, the children living in homes in which remediation was actively carried out had fewer days in which their asthma was symptomatic than did the control group. This effect held both during the study and afterward. Just 1 in 29 of the kids in the remediation group saw exacerbations of asthma symptoms after the study. In the control group, 11 out of 33 kids did. Hooray, a simple solution![38] The study was small, it was in just one city, and yet it is hopeful in terms of suggesting a way forward.

For now, what we can say is that if your house gets wet, you should figure out how to fix the water problem and dry it out. If you are building a new house, you might avoid drywall, particularly in areas that get wet, because you can't be sure that it doesn't already have *Stachybotrys chartarum* in it. And if an opportunity comes up to help support research on the biology of fungi in houses, sign up. Meanwhile, the fungi on the ISS continue to thrive, a reminder that whatever the right solution is to managing the fungi in our homes, it is, just as for bacteria, very unlikely to lead to their eradication. This is something NASA scientists, the Russians, and Birgitte can agree on.

As for the tens of thousands of other fungal species we have found in homes, each with a story as elaborate as that of *Stachybotrys chartarum*, they need to be studied. You are breathing them in now, these poorly understood species. Thousands of them are so unfamiliar they do not even yet have names. You could be the one to name them. You're right to be skeptical of the idea that thousands of species around you are not yet named, but it is true. To some extent, this just reflects our broader ignorance about Earth in general. We've only begun to explore the planet. Most of life is not yet named. With bacteria, we haven't even scratched the surface. For fungi, we are, perhaps, a third done with naming and far less completely finished with what comes next: the study of the details of each species' biology. For insects, we might, if we are lucky, be half done. But I think there is also something specific at play in homes. In our homes, we tend to study species we know pose danger to humans, but there is no one assigned to study the rest of the species. Basic biologists might study them. But given the choice, most basic biologists would rather go off on trails in the woods, exploring remote locales (such as field stations in Costa Rica). We have blinders on blocking our view of the wildlife that is innocuous and close at hand, a reality that became very clear to me recently when we asked people about what lived in their basements.

# 7

# THE FARSIGHTED ECOLOGIST

Of the animals that live with men there are great numbers. . . .

—HERODOTUS

A little wind carries the ship. A little bee brings the honey. A little ant carries the crumb.

—ONE TRANSLATION OF A SECTION OF THE INSINGER PAPYRUS XXV.1 TO XXV.4

And there came a grievous swarm of flies into the house of Pharaoh, and into his servants' houses, and into all the land of Egypt: the land was corrupted by reason of the swarm of flies.

—EXODUS 8:24

WE HAVE MISSED seeing and understanding the bacteria and fungi in our houses, and their consequences, in part because they are small. When it comes to animals, though, something else is going on. I've come to believe there are reasons why ecologists and evolutionary biologists have failed to pay attention to the animals in our houses even though such species are larger.

Ecologists are professionally far sighted; they see species in remote locales more clearly than those closer at hand. Farsightedness sounds like a good thing, but it isn't when it means missing what is most immediate. In New York City, for example, scientists have taken many samples of animals in the forests surrounding the city, but far fewer in the city itself. Fewer still indoors. This isn't accidental. As ecologists, we're trained to study life in "nature," which we have come to believe means the absence of humans. This bias is even built in to our most important surveys of animal life. The breeding bird survey, for instance, the biggest structured survey of birds in North America, excludes parts of the United States that are heavily urbanized. It excludes the places we live. Ecologists, as a result, have good data about the exact location of the rarest of North America's birds, but not about the abundance of house sparrows, pigeons, or crows. The same is true with insects, only more so, as I became very aware of when I began to study camel crickets.

Humans have been living alongside camel crickets for a long time. When our early ancestors stayed in caves, they inevitably encountered other animals there too. We know of these encounters because of the bones we find in caves and the claw marks on cave walls, but also because of the species we find depicted in cave art. Some of the animals in caves were big and dangerous. Imagine climbing deep into a dark, damp tunnel, your travels illuminated only by a stick red with embers. Imagine then seeing, perhaps smelling first, a cave bear. Cave bears (*Ursus spelaeus*) could be as large as the very biggest grizzly bears. When the fates were willing, our ancestors killed cave bears. When the fates weren't so kind, they were killed by them.[1] But in addition to cave bears, our ancestors also encountered smaller species. These probably included bed bugs and lice. They definitely included camel crickets, which we know on the basis of a single carving.

The cave in which the carving was found was discovered by three boys. In 1912, Max Bégouën and his two brothers, Jacques and Louis, heard about a place where a small stream on their property in the French Pyrenees went underground. Their neighbor,

Francois Camel, suggested the boys follow the stream deeper into the Earth. So they did. In doing so, they discovered one underground chamber after another until their path was blocked by stalactites. This was, of course, the stuff of childhood dreams, but it was also the end of the path. Then one of the boys spotted a narrow hole in the stalactites, high up in one chamber. The hole was the width of a boy. Max and his two brothers squeezed in the hole and continued along a passageway. Deep within the passageway, they climbed up a forty-foot chimney of stone. At the top of the chimney, they reached another chamber. The chamber—a room, really—was filled with the bones of cave bears. Among the bones were two well-rendered statues of clay bison.

Two years later the boys stumbled upon more of the cave. In 1914, on the other side of the hill, they found a hole in the ground. They lowered themselves in and discovered an eight-hundred-meter-long cavern. The boys explored the cavern and then crawled through a narrow tunnel at one side, where they found another chamber. In the chamber, the boys found themselves face to face with one of the great masterpieces of cave art: a shaman, part-human, part-animal, emblazoned with antlers. On another wall in the same chamber, teeth, charcoal, and bones were shoved like votive pieces into the clay beneath a carving of a lion.

One particular piece of bone in the cave (which would come to be named Trois-Frères, "three brothers," in honor of the boys) bore a unique carving: a camel cricket[2] of the genus *Troglophilus*. This depiction was evidence that our ancestors (or at least one of our ancestors) paid attention to such animals. Over the next ten thousand years, many humans would come into contact with camel crickets, in caves and also in houses.[3] In the basements and cellars of homes, we have re-created conditions similar to those of caves, conditions that provide some camel cricket species with just what they need. Off and on, then, we have associated with camel crickets longer than we have farmed crops. Though our history of association with camel crickets is ancient, and camel crickets can reach extraordinary abundances, they have not been very well

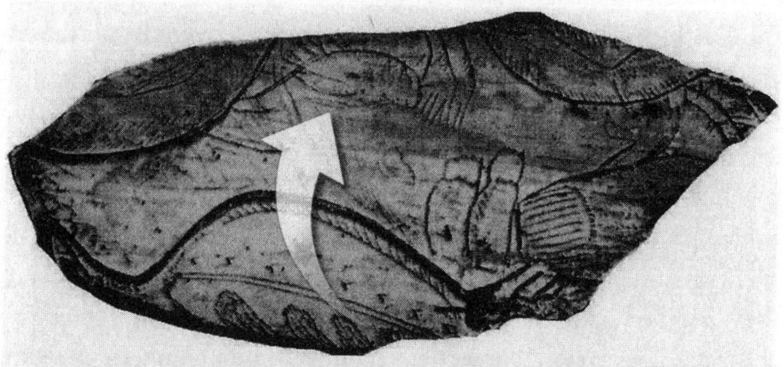

**Figure 7.1** A bison bone fragment bearing a carving that clearly depicts a camel cricket of the genus *Troglophilus*. The carving was discovered in the Trois-Frères cave in the central Pyrenees and is one of the only depictions of an insect in European cave art. *(Image modified from an original drawing by Amy Awai-Barber, originally published in* Cave Biology: Life in Darkness, *by Dr. Aldemaro Romero.)*

studied. In my studies of camel crickets, they have come to seem like an emblem of just how easy it is to fail to notice the species around us, especially when they are in plain view.

Camel crickets have been interesting to me since I read Sue Hubbell's book *Broadsides from the Other Orders*[4] when I was an undergraduate. Hubbell, a writer with no real scientific training, kept camel crickets in a terrarium. She was a patient and curious observer, which was more than enough: she made discovery after discovery about the camel crickets' biology. Some of what she discovered stuck with me, but what I remember best is how much, after her years studying those crickets, was still unknown. Those unknowns included very basic things, such as what camel crickets eat.

My lab members and I decided to do pick-up where Hubbell left off, beginning with a really simple study: a census. Already connected to thousands of participants across our various projects, we asked them whether they had camel crickets in their basements or cellars. Within a year and a half, we got 2,269 responses, enabling us to map where basement camel crickets live. There was a big surprise. The map contradicted much of what we thought we knew about the insects' distributions.

Many of the camel crickets native to North America are of the genus *Ceuthophilus*, a genus that includes eighty-four species (this is the number so far; more will probably be discovered). At some point, as Western-style houses spread across North America, populations of *Ceuthophilus* camel crickets moved in. In the wild, most camel cricket species live in caves and dark places in forests, such as leaf litter. They live a hard life of jumping, bumping, and subsisting. Long antennae enable them to sense odors, the cold, and humidity. Adapted to life in the dark, they have tiny eyes that look, as Sue Hubbell noted, like small buttons. In the wild, it is supposed (though not known) that they eat the bits and pieces of low-nutrient food that float into caves or sift onto the forest floor, dead things and long-dead things. If so, that would make them key links in food webs, especially those in caves, because they can subsist on food that few other species eat, such as carbon compounds that are too hard for other organisms to break down; the camel crickets are then food for other life.[5] Camel crickets in houses probably play a similar role, making inedible bits of your basement edible for spiders and mice.

Not all North American species of camel crickets have moved into our homes (and some are still very restricted to living in caves and should probably be listed as endangered), but at least six have. The distributions of the six known home-dwelling camel cricket species were studied in the early 1900s by Theodore Huntington Hubbell at the University of Michigan. Hubbell was, along with a student of his, Ted Cohn, one of just a handful of people to focus attention on camel crickets. He wrote the book on these bouncing beasties, a five-hundred-some-page treatise called *The Monographic Revision of the Genus Ceuthophilus*. Although it deals with the evolution, geography, and natural history of crickets, it reads a bit like the Old Testament—the story of who lived where and who begat whom. It's not particularly fun to read, except perhaps for the obsessed, but it was critical to our work. Hubbell's book made

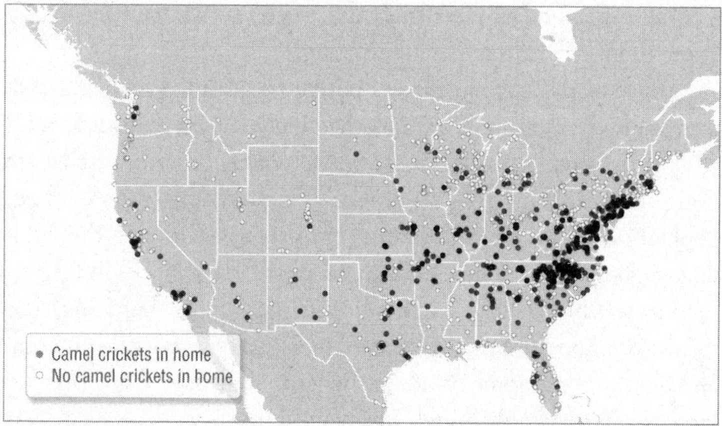


**Figure 7.2** A map of the households that reported whether they had or did not have camel crickets in response to our email questionnaire. *(Credit Lauren M. Nichols, data from MJ Epps, H. L. Menninger, N. LaSala, and R. R. Dunn, "Too Big to Be Noticed: Cryptic Invasion of Asian Camel Crickets in North American Houses,"* PeerJ *[2014]: e523.)*

clear that we should expect camel crickets to be found both outside and inside homes across North America, with absences only in the very coldest places. One or another species of camel cricket could be found, at least occasionally, basically everywhere. This meant that if we were to map the distribution of camel crickets in homes we should see a broad swath of camel crickets across the continent with both presences and absences in every region. We did not (see Figure 7.2). Instead, camel crickets were common in basements in eastern North America but seemingly rare or even absent in much of northwestern North America. Something seemed wrong.

One possible explanation was that our public participants were not very good at studying their own houses. Maybe they were mistaking cockroaches for camel crickets or vice versa, or people in the Northwest were too scared to look, or some regions just have so few basements that little habitat is available to camel crickets. Maybe it was a combination of all of these things. It turned out to be none of them.

At precisely this time, MJ Epps joined my lab as a postdoctoral researcher. MJ—Mary Jane, although I suppose she is only ever called by her full name when in trouble with her mother—is an extraordinarily talented natural historian and ecologist. MJ knows beetles. MJ knows fungi. MJ knows the woods.[6] The camel cricket mystery seemed like a great project for her to get started on. I asked her if she could figure out what was going on with the distribution of camel crickets. MJ, working with Lea Shell, whose job at the time was to help us effectively engage the public, asked people to take pictures of the "camel crickets" going bump in the night in their basements.

Between January 2012 and October 2013, we received pictures from 164 homes. Some pictures showed dozens of camel crickets dead on sticky traps. Other photos showed things we couldn't decipher. But 88 percent showed the exact same thing: a surprise. An answer we hadn't expected at all. In those photos we saw one or more individuals of *Diestrammena asynamora*, a giant species of Japanese camel cricket known to be in the United States, but not really known to be in homes. We finally had an explanation for our map of camel crickets. It didn't match our understanding of where native camel crickets lived because it wasn't a map of the native species at all. We had mapped an introduced species whose distribution did not correspond to the old maps because it had moved in since the maps were made!

Based on what we were able to find in museum collections of insects and old reports and papers, it appears that the Japanese camel crickets made the journey from Asia at least a hundred years ago. Many species from temperate Japan or China have been introduced to the United States. Such species tend to be described as Japanese, in part because they are often better studied in Japan than they are in China. We can and eventually will study the genetics of these camel crickets to better understand where they are native and when and how they arrived in North America. We haven't yet, though, so it is difficult to reconstruct the details of their movements across North America. It appears that for much

**Figure 7.3** A camel cricket, *Diestrammena asynamora,* in a Boston basement. *(Photograph by Piotr Naskrecki.)*

of their history in the United States, these crickets confined themselves to greenhouses (and the occasional outhouse too). More recently, they made the move into houses. Once in houses, the crickets were probably seen by many people, including thousands of scientists. Their invasion was overlooked in plain view. Just what allowed the Japanese camel crickets to make the shift indoors is not known. Possibly the crickets evolved new traits that enabled them to thrive in the drier, cooler conditions of houses. Alternatively, it may simply have taken time for the camel crickets to hop basement to basement across the country.

Nor was this *Diestrammena asynamora* species alone. When we looked more closely at the pictures, we found that some of the camel crickets were actually of a second species of *Diestrammena, D. japonica,* also apparently from Japan.

Upon discovering the identity of the most common camel crickets in homes, MJ wanted to figure out just how many there were. She worked with Nathan LaSala, then a high school student, to sample camel cricket populations in ten houses in my neighborhood in Raleigh. Nathan's job was to set traps (plastic cups of the sort that college students use to play beer pong) at increasing distances from houses where camel crickets were known to be living. The idea was that we could then estimate how far camel crickets fan out from houses. We hoped that college students wouldn't just take the cups or, worse yet—and I'm not sure why this was the specific worry, though somehow we all hit upon it—pee in the cups. Assuming the cups were left alone, the trick was figuring out how to trap camel crickets. I didn't have a clue. But MJ did. Like a modern Pippi Longstocking (with an Appalachian accent rather than a Swedish one), she smiled big, laughed, and said, "Camel crickets are easy to trap with molasses. Everyone knows that!" Well, whoever "everyone" might include, it wasn't me. But MJ was right. She had Nathan bait the traps with molasses and, sure enough, they caught camel crickets, though fewer and fewer the farther the traps were from houses. Nathan and MJ used these results, along with the estimates of the proportion of houses hosting camel crickets, to extrapolate the total number of giant, Asian, introduced camel crickets (*Diestrammena asynamora*) that might be living in eastern North America (assuming the biology of the camel crickets is representative of their biology more generally, which so far seems to be the case). The number, which I suspect is conservative, was a whopping seven hundred million camel crickets—nearly a billion thumb-sized animals living in our houses, unnoticed.

Here, then, was a crazy moment. Not one but two species of relatively giant insects had moved in right under our noses. What does this mean about our ability to detect the vast majority of much smaller species and their movement? I wasn't sure, but it seemed to suggest we might be missing them too. MJ went on to write a scientific paper about the camel crickets. The discovery was, to us, a big deal. We'd been living with a totally unstudied species—a large unstudied species—for years (perhaps decades) without realizing it. We felt like the Bégouën brothers except that the cave full of wonders into which we had stumbled was a basement. And so, like the Bégouën brothers, we continued to explore.

THE DISCOVERY THAT almost a billion thumb-sized Japanese camel crickets were living in houses without anyone really knowing they were present left me a bit dumbfounded. It was easy enough to figure out what had happened. If you aren't a scientist and you see a camel cricket in your house, you assume scientists know what it is. If you are a scientist, but not an entomologist, and you see a camel cricket in your house, you assume that entomologists know what it is. If you are an entomologist and you see a camel cricket in your house, you assume the specialists in camel crickets know what it is. Meanwhile, just two people on Earth specialize in the study of camel crickets and neither of them happens to live in a house where the Japanese species is present. I started to wonder whether this phenomenon—of assuming someone else knows—is likely to be more common in homes than in other habitats, more common because homes are the place we are most likely to assume that someone else knows, most likely to assume that everything is more or less under control. If I was right, it meant that not only was the home a place where it was still possible to make new discoveries but also it might be an ideal place to make discoveries, discoveries that, because they implicitly affect many people, would be important.

The trick was how to test such an idea, what I call the "far-sighted ecologist syndrome." I could look in museum collections and see where the specimens in those collections tended to come from. I eventually did. What I found was that entomologists are, indeed, less likely to collect specimens in the places where people actually live. And even when they do collect insects in the places people live, their focus tends to be on particular species. Nearly all of the collections in Manhattan over the last twenty years, for example, are from Central Park, and even those collections tend to be of just a handful of species, primarily honeybees, aphids, and soil mites. But this might just be because little lives in the parts of

Manhattan where people are most numerous. I was talking about this one evening at the house of my friends Michelle Trautwein and her husband, Ari Lit. We were visiting for a long dinner. Michelle and Ari are my close friends. Michelle also happens to be a world expert on the evolution of flies. At the time, we had just begun to unravel the story of the camel crickets. In its light, Michelle and I pondered what could be found if we exhaustively sampled the arthropods in houses, whether in Raleigh or in New York. We walked from one window sill to another, glasses of wine in hand, and looked at the insects present. We saw several kinds of spiders, some drain flies, and even a couple of species of beetles. Neither of us knew all the species, but it was easy to imagine someone else did. Maybe we were suffering from farsighted ecologist syndrome ourselves! We could check to see what the species were. Or better yet, we could sample a bunch of houses to see what has been missed. We thought it might be a lot, maybe even hundreds of species. It was late enough in the evening that these insects were sufficient inspiration to cause us to ponder both the limits of the universe and the possibility of a new research project. The timing was good: Michelle was starting her own research program at the North Carolina Museum of Natural Sciences. We would sample the arthropods of homes and figure out what was present. We toasted to insects and returned to our spouses at the dinner table to talk about the world.

Not everything that seems like a great idea while sipping drinks is still a great idea in the morning. The next day, though the idea still seemed good, problems arose. For one, each and every entomologist we talked to about it seemed to think it was boring. Graduate students we tried to enlist as assistants were more interested in studying remote forests and mostly declined our invitations. I called a friend who suggested that if I wanted to find a bunch of species I should just go find a log in the rain forest and pick it apart. "Don't waste your time on windowsills and kitchens, man. Let's go back to Bolivia!" When Michelle and I were feeling optimistic, we thought everyone was wrong. But in other moments, we wondered whether they were right. Maybe the camel cricket was just a strange exception. Still, we would go ahead.

One potential trick to carrying out the project was identifying the species we found. I could identify the ants. Michelle is a fly specialist (she is now the curator of flies at the California Academy of Sciences) and could identify a subset of the flies. We would just beg and borrow help identifying whatever else we found. How much more could there be? Still, just in case we found any really-hard-to-identify species, we enlisted Matt Bertone. Matt is an entomologist's entomologist; he is unusually talented at identifying insects, a task that pleases him so long as he can do it slowly and deliberately, on his own schedule. He agreed to our plan, though not before voicing the opinion that we wouldn't find very much. As the plan developed, we added other people to the research team, each with their own specialization and skills. Inasmuch as volunteers were not forthcoming, we paid the whole team to go house to house, catching, counting, sorting, and identifying bugs. It was probably overkill. We were probably overthinking the project. How many species could we possibly find in homes anyway? I had a dream about the project. In it, we had sampled ten houses and found in them nothing but six cockroach legs, a kid's pet praying mantis, and a louse the size of a rabbit that none of us could catch. It was a dream that was both inauspicious and strange.

When the team entered houses, it was burdened with bug-collecting gear. Jars. Nets. Notebooks. Aspirators. Hand lenses. A portable microscope. Cameras. It looked like some kind of entomological circus in which the only missing pieces were a fire-eater and some pounding drum music to set the mood.[7] If we were successful in finding interesting species, this parade of people and devices would be a grand beginning. Implicitly, if we were unsuccessful, it would just be silly and grandiose.

I was in Denmark at the time, with my family, trying to convince the Danish Natural History Museum to do a similar project with houses in Denmark. It didn't work; no one thought we would find anything. Perhaps as comeuppance for my absence in Raleigh,

**Figure 7.4** Matthew Bertone, entomologist and insect identification guru, collecting arthropods in the nooks and crannies of a home (while simultaneously taking a photograph). *(Photograph by Matthew A. Bertone.)*

it was decided that the very first house to be sampled would be my house there. Matt, Michelle, and the rest of the team trudged up my front steps. They would, subsequently, enter another forty-nine houses in Raleigh followed by additional houses around the world.

In each house, including mine, the team searched room to room. Some searches took as long as seven hours. Often, houses looked as though they might contain few insects, but life was found in nearly every room. It was in the corners. Or in the drains. The team didn't quite go so far as to flip the pages of books looking for life, but nearly so. Windowsills were mortuaries of dead insects as were light fixtures. The spaces under beds and behind toilets often offered up discoveries as well (some unwelcome). Each time the team found an arthropod, living or dead, they put it in a vial or jar. Homeowners looked on with surprise as jars that started empty filled with transparent ethanol browned with ever more legs, wings, and bodies. The brown jars were a good sign, or seemed to be anyway (at least to us; the owners' sentiments may well have been more complex). Tallying this life and identifying just what had been found would be done back in the lab, and it would take many months. Because different people were collecting in different rooms and no one saw everything that was found, it was hard to get a sense of the full picture.

When I'd write Michelle from Denmark to ask about the status of the collections, she'd remind me that identification takes time, that Matt works better when he can do things carefully. Michelle told me to be patient (knowing me well enough to know that was unlikely). Also, she said there seemed to be more specimens than we anticipated (there were, in fact, more than ten thousand specimens, though no one knew it then). Each specimen, however small or partial, had to be pulled out of its jar, labeled, and identified individually. To do this identification work, Matt looked not at the entire body of each insect but instead at those particular features that distinguish one species or genus from another. The important characteristics are different for each kind of insect. Some ants are distinguished based on their number of antennal segments, whereas really knowing a click beetle requires a careful inspection of the hairiness and shape of the penis.[8] Sometimes, even that wasn't enough to go on, and Matt had to send a specimen to a systematist specialized in, say, drain flies. The specialist might live in Ohio, Slovakia, or New Zealand, so the specimens were shipped, which took even more time. Many groups of insects are known in detail by just one specialist in one place in the world. In such cases, Matt labeled the specimens nicely, packed them well, and shipped them for further identification, which might take weeks or, if the specialist was very busy, decades (some of our specimens still await identification). Many a systematist, in contemplating his or her own death, fears dying surrounded by specimens in boxes, great piles of specimens, that they never had a chance to identify.[9]

Finally, the first house, my house, was done. My house contained no fewer than one hundred species of arthropods. I say "no fewer than" because some insects couldn't, and still can't, be identified, either because there are no experts to identify them or because they were in sorry shape (a dried-out wing, a pair of disembodied legs, a single compound eye). One hundred! That was crazy—ten or even twenty times more than most entomologists had guessed we'd find. What was even more amazing was that the first house, my house, was not unusual. Nearly all of the houses sampled contained at least a hundred species of arthropods (and more than sixty families of arthropods). Some had many more, up to two hundred species. Nor was Raleigh unusual. Over the next years, we would find a similar diversity in houses in San Francisco and Sweden and in our more extensive (but less intensive) survey in which we considered which arthropods were present in houses based on the DNA in house dust.[10] Houses in Peru, Japan, and Australia proved even more diverse. We have found thousands of species of arthropods in homes. In Raleigh alone, these arthropod species came from 304 arthropod families. A family is a taxonomic unit larger and more ancient than a genus (a genus is older than a species, a subfamily is older than a genus, a family is older than a subfamily). All ants, for example, represent a single subfamily, Formicidae. In houses we found more than three hundred families of arthropods as unique and old as the ants. An entire world of animal life in plain view had been missed. They were not missed because they were microscopic. They were missed because they were overlooked, invisible in plain view. Look around now. No matter how tightly sealed your house or apartment is, there are arthropod species inside with you. Look carefully—they are there. We promise. You can stop reading and go hunt for yourself if you want. I'd start with the windowsills and light fixtures.

The obvious next question was which species of arthropods we were finding. Lots of flies, hundreds of species of flies in total, quite a few of them likely to be new to science. House flies, fruit flies, scuttle flies, nonbiting midges, biting midges, mosquitoes, lesser house flies, phantom midges, freeloader flies, and shore flies. This is not to mention fungus gnats, moth flies, and flesh flies. Or crane flies, winter crane flies, and minute black scavenger flies. Or long-legged flies. Or dung flies. If you see two flies in your home, the odds are that they are two different species. Heck, if you see ten flies in your house, they are likely to be *five* different species. The next most diverse group was the spiders (house spiders, wolf spiders, ghost spiders, jumping spiders, spiders that spit venom on their prey, and many more), followed by the beetles, then the ants, wasps, bees, and their kin. Even the millipedes were diverse. We found millipedes from five different families in homes. Aphids were also common in houses in Raleigh as were the wasps that lay their eggs in the bodies of aphids as well as the wasps that lay their eggs in the bodies of the wasps that lay their eggs in the bodies of aphids.[11] So too were wasps that lay their eggs in the bodies of cockroaches, tiny wasps unable to sting you but perfectly prepared to sink their sting-like ovipositor into the egg case of a cockroach and, having done so, to plant their eggs beside the baby cockroaches they will hatch and eat. To look at all of this diversity, you come to see, as Annie Dillard wrote, "the curious shapes soft proteins can take." In considering these shapes, I came to "impress myself with their reality and to greet them," to greet them as roommates, mine and yours.

At first, entomologists told us that we wouldn't find many species in homes. When we found thousands of species, they then argued that these had all just drifted in. Houses, they said, were giant light traps, and so simply caught what was otherwise around. During a talk one of us gave, a colleague said, "What does it even mean that those species are there? They are doing nothing." Academics can be real ninjas of passive-aggressive combat. The trick was figuring out how to see which species were really a persistent part of the story of the house. One approach, the one we took first, was to try to understand which of these thousands of species were found in homes not just occasionally but over months, weeks, and years. We redoubled our efforts to find studies from other regions with which we could compare our results. We found nothing. Then, finally, we found two other studies. The first was of chicken barns in the Ukraine. The Ukrainian study focused on spiders and the species trapped in spider webs. Of the seven most common

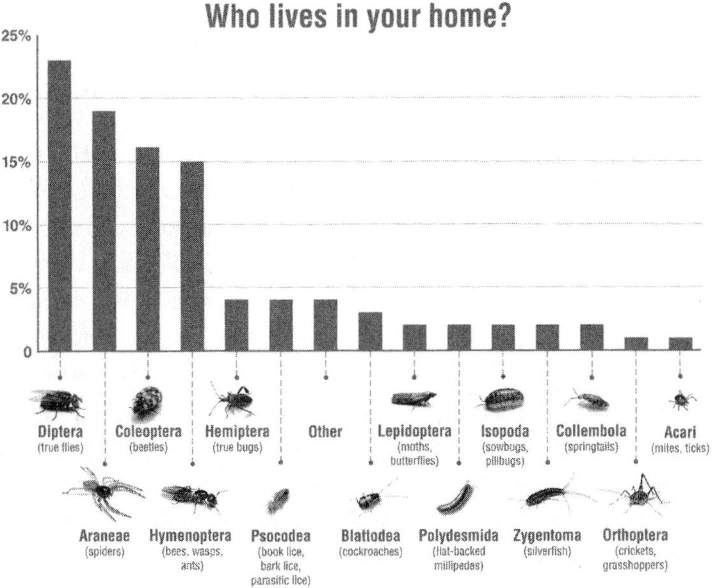


**Figure 7.5** The proportion of species in Raleigh houses from different arthropod orders. *(Modified from a figure by Matthew A. Bertone.)*

spider species in the chicken barns in the Ukraine at least four were also found in houses in Raleigh (the study couldn't find anyone able to identify the insects), suggesting these species may well be predictable, global, indoor dwellers. The second study was work done by the archaeologist Eva Panagiotakopulu.

Eva is a special sort of archaeologist, one focused on the insects of ancient houses. Some people yearn to be a fly on the wall. Eva and her colleagues just want to know whether there *was* a fly on the wall. Eva has studied the arthropods in homes in ancient Egypt, Greece, England, and Greenland. In doing so, she has revealed a picture of the species that live with humans and how they have moved around the world. She can't study all of the arthropods in those ancient homes, just those that belong to the few families that tend to preserve well either as adults (for example, beetles) or as pupae (for example, flies). Eva has a narrower window into ancient life than we do into modern life, but it is a window that allows her to look across great spaces and times.

The species Eva and colleagues tend to find in ancient homes around the world are typically associated with food (beetles that eat grain, beetles that eat flour, beetles that eat the fungus that grows on grain and flour), waste (dung and carrion beetles), and other corners and habits of daily human life. Nearly all of the tens of grain- and food-associated arthropod species Eva has found in ancient homes (in Amarna, Egypt, for instance, from 1350 BCE) are also found in Raleigh. So too were many of the waste- or body-associated species. Each species that has been studied in any detail is unique, yet the patterns repeat. Species moved from wild nature into human homes and found food resources. They were then inadvertently carried in human food, on building materials, or even on human bodies from place to place. This is the story of house flies, fruit flies, Indian meal moths, some dermestid beetles, and even some cockroach species. In the Bible, Noah filled the ark with lions, tigers, and the like. On the real ark of our human peregrinations, we brought two of each of many kinds of insects instead. It didn't take very long for the insects to follow us to different continents.[12] An outhouse in Boston, dating from the year 1650, contained a bowling ball, porcelain, shoes, and no fewer than nineteen species of household beetles that had already been introduced from Europe.[13]

By comparing our results for homes in Raleigh to the work done by Eva and her colleagues, we estimate that no fewer than a hundred and perhaps as many as three hundred species of arthropods made the journey all the way from the Near East or Africa into homes in Raleigh (and nearly everywhere else in North America). Some other species in homes in Raleigh may be those that associated with Native American homes before colonists arrived, including a few species of carpet beetles. Other species have had more unusual travels. Human fleas, for example, appear to have evolved on guinea pigs and then somehow traveled among human populations from the Andes all the way back to the Near East and Europe, perhaps on furs that were traded.[14] In short, homes include hundreds of species that have lived in homes long enough to evolve specific adaptations for indoor dwelling, species that have, whether or not we pay attention to them, been a far more predictable feature of our history than, say, democracy, plumbing, or literature.

In addition to the species in homes that specialize in living indoors (and which we might expect to find in dwellings in different regions as well as in dwellings from different time periods), we found hundreds of species in homes that really are outdoor species. Some outdoor species come in mostly to eat, such as the thief ant, *Solenopsis molesta*. Others, such as the smallest cricket in the world, a species of the genus *Myrmecophilus*, just follow the species on which they depend. *Myrmecophilus* crickets live with ants and were present in some of the houses with ants. Similarly, Matt Bertone found a larva of the bearded lacewing in one house that also had termites; this species is rare and lives in termite nests, where it releases a "vapor phase toxicant" from its anus[15] that it aims at termites to stun them, several at a time, before eating them. Nature can be ridiculous. Other outdoor species we find indoors have just lost their way, such as the many species of aphids, for example, or the wasps that lay their eggs in aphids, or the wasps that lay their eggs in the wasps that lay their eggs in aphids. We also found honeybees and bumble bees and solitary bees indoors. These species flew in accidentally, but they still tell us something about homes and our lives. They are a measure of the biodiversity in backyards, the biodiversity both of the insects and of all the plants and other species on which they depend. Or, where such backyard biodiversity was absent, they failed to fly in as a measure of its absence.

For the vast majority of the arthropod species we find in homes, we don't know what they eat. We don't know where they are native. We don't know what their closest relatives are. When you see them in your kitchen, it is not so different from when I, as a twenty-year-old, saw insects beneath leaves in the rain forest in Costa Rica. In Costa Rica, it is a safe assumption that what you see beneath a leaf has rarely or even never been studied and that anything you might notice about its biology is new to science. For the species in homes, it increasingly appears that the same is true too. With one difference. Many thousands of scientists, and millions of people more generally, are likely to have seen the species you see in your house. They have just failed to pay much attention. Recently, a study found thirty new species of phorid flies in urban Los Angeles.[16] Then the authors of the study kept working and found another twelve new species in Los Angeles.[17] On the other side of the continent, New York City has also, recently, proven a font of new species. A new species of leopard frog was found in the city (*Rana kauffeldi*), then a new species of bee (*Lasioglossum gotham*) and a dwarf centipede (*Nannarrup hoffmani*).[18] Then a new species of fly.[19] These studies focus on the outdoors rather than the indoors and yet they echo my main point. We wake amid the unknown even when it comes to species that are visible, perhaps especially when it comes to such species. I suspect quite a few of the arthropod species we found in homes are also new species, but to be sure, each one has to be identified not just by Matt Bertone but also by the expert on that particular group of insects—the moth fly expert or the stone centipede expert. And as often as not, there are no experts.

What I've taken away from our work, so far, with the arthropods in homes is that when you see a species in your home, you should study it. You should pay attention. Don't assume someone else has already figured everything out. Take pictures. Make drawings. Pull out your hand lens and a notebook and record what you see. Then, if you see something interesting, do what Leeuwenhoek would have done: use the tools you have to figure out what it is and what it might be doing. Then send a letter to a scientist. The tools for identifying the species in your home are better than they have ever been; so too the ways of connecting what you have found with scientists. Antony van Leeuwenhoek, one man, working alone, discovered new species and phenomena virtually every day. Imagine what we could all do working together. We don't even know the simplest things, such as which species eat which other species in our houses. Record what the spiders in the corners of your house catch. Or you could catch an arthropod and keep it in a terrarium and watch what it feeds upon or how it mates (it was by doing this that the science writer Sue Hubbell documented features of the love lives of daddy long-legs that no scientist had ever before seen). I'm ever more convinced that not only do discoveries lurk among the animals in our homes but also they are especially likely there. But even now, even after all our work, even after making discovery after discovery in homes, when I mentioned this to Michelle Trautwein, she said, "But can you be sure that it isn't just that discoveries are everywhere and we happen to be studying the home?" I guess I still can't be sure, but perhaps that is the bigger point. We know so little even about the animals around us that we can't preclude the possibility that some of the biggest discoveries yet to be made are those right where we wake up every morning.

Many entomologists imagined our homes would contain few insect species, most of which would be problem pests. Yet real problem species, such as house flies that vector fecal-oral pathogens, German cockroaches that trigger allergies, termites that eat houses, and bed bugs that make us itch, were actually rare in the houses we studied. Instead of these species, we, like the Bégouën brothers in their cave, found that the chambers into which we stumbled were full of mysteries. We found a diversity of small animals, a beautiful and sublime manifestation of the ancient history of animal life.

Yes, I find aesthetic value in the species of animals that run around our homes. You don't have to agree with me. I can't make you. But why might you? Why might you, when the species we are talking about are precisely the same species that many adults find bothersome or even vulgar? In considering this question, I'm reminded of an essay in a book written by Harry Greene, the snake biologist and natural historian.[20] Greene was considering a similar question for snakes. Greene, building on the work of the philosopher Immanuel Kant,[21] distinguishes between two aesthetic values nature (be it snakes, spiders, or anything else) can have: nature can be beautiful and it can be sublime. Beauty is what we experience when we see the colors of an individual cardinal, listen to the song of an individual chickadee, or watch a singular whale rise out of the water. Beauty is an experience influenced by our senses and our culture, but not by intellectual context. The other day, looking through a microscope, I saw the scales on the wings of an Indian meal moth and found them beautiful. So too the web of a house spider hanging above my front door and even the antennae of a mosquito. The sublime, though, is something different; it is an aesthetic appreciation that goes beyond our observation of an individual insect or bird and that instead contextualizes observations in light of a broader understanding. The sky can be beautiful inasmuch as the pattern of the stars is visually appealing, but its sublimity results from our awareness of the grandness of the universe and the reality that each prick of light is a star no greater or lesser than our sun. It was the beauty of the cave in France that first impressed the three Bégouën brothers, but it was the sublimity of understanding its place in the early history of human artists that led the brothers, and particularly Louis, to dedicate much of the rest of their lives to the study of the cave. Similarly, while the wing of the Indian meal moth is beautiful, understanding that the meal moth is the same species that probably lived on one if not all of Columbus's ships, the same species that would have flown up out of grain in ancient Rome, and the very same species that was likely to be present in ancient Egypt is sublime. It is also sublime to know that similar stories could be understood and unraveled for each of the species in our homes, but that they haven't been yet. I find the unknown and the yet-to-be-discovered as raw and thrilling as is the magnitude of the universe. I have felt that way since I first wandered down a trail in a Costa Rican rain forest. To me, the beauty and sublimity of the species of arthropods in our homes, just like that of arthropods anywhere else, is reason enough to care about them, watch them, and even, in some cases, conserve them. But you may find you still disagree; you may find yourself wondering what good all of this life really is to you. If you are, you are not the only one.

# 8

# WHAT GOOD IS A CAMEL CRICKET?

Don't worry, spiders,
I keep house
Casually.

—KOBAYASHI ISSA,
*The Essential Haiku: Versions of Bashō, Buson, and Issa,*
translated by Robert Hass

WHEN MY COLLEAGUES and lab members and I started to write papers about the camel crickets and other arthropod species in homes, we were incredibly excited. We'd found so many species that we could imagine having hundreds of students work on them for decades. In our enthusiasm we expected that when we wrote scientific articles about our discoveries and then shared them with the public, the public would be excited too. We imagined inspiring thousands of eight-year-old girls and boys to go forth in their own homes to study life-forms no one had ever studied before. To some extent this happened. I hope it continues to happen, and a large part of what my lab now focuses on is finding ways to make it easier for kids and families to help study more about the species around them. But enthusiasm was not the only response to our discoveries. Some people also asked, "Well, how do I get rid of them?" or, even more often, "What good are they?"

To ecologists, when someone asks what good a species is, it chafes. It is a weeping foot fungus of a question. As ecologists, we learn that any particular species isn't good or bad, nor of greater or lesser intrinsic value than any other—they simply exist. Except in the light of our own beliefs and wants, a blue whale has no more value than does the tapeworm in the blue whale or the bacterium in the tapeworm or the virus in the bacterium. They exist simply because they evolved. The same for the genital louse or the human botfly, a species whose larvae live in human subcutaneous tissue and breathe through two snorkel-like spiracles. No goods, no bads, just existence.

But just because we wouldn't ask a question (or wouldn't ask it a certain way) doesn't mean that there isn't some version of the question that isn't very interesting. Rather than simply dismissing this question, we can reframe it as "What use might this species have to human society, and how could you use ecology and evolutionary biology to figure that out?" The change is subtle (and a bit wordy), but it turns the question into something scientists are better able to think about. A variety of species from houses have indeed proven useful to humans.

I've already talked about the value species in our homes can have for us directly in terms of our health and well-being. But many species can also indirectly benefit us via the effects they have on particular industries. Mill moths (*Ephestia kuehniella*), for example, which can be abundant in kitchens and bakeries, are afflicted by a pathogen called *Bacillus thuringiensis*. This pathogen was first found in mill moths in Germany (in Thuringia, to be specific). It was later discovered that this pathogen could be used to kill pests on crops. The live bacteria can be sprayed on organic crops. Then it was discovered that the genes from *Bacillus thuringiensis* could be inserted into the genomes of corn, cotton, and soybeans. Transgenic versions of these crops now produce their own pesticides. The mill moth was useful because it was host to a bacterium that turned out to have genes that sustain a multi-billion-dollar agricultural innovation.

Dozens of species of *Penicillium* fungus can be found in houses. It was one of these species in which antibiotics were first discovered, a discovery that ultimately saved many millions of lives. Another *Penicillium* species was the source of the first cholesterol-lowering drug (a statin). House mice and Norway rats are both household species, species that became common by taking advantage of houses. Together with the fruit fly, house mice and rats are the animals we use to understand the workings of bodies and medicines so that we don't have to experiment on humans. Fruit flies, house mice, and rats are "good" because they have allowed us to perform medical research without hurting humans. We understand ourselves by studying them.

I can list many more examples, but as I thought about the uses of the animal species in houses it occurred to me that perhaps I could do better than just make a list. Maybe I could actually begin to systematically search for the uses of different household species in the lab. I would start with the introduced camel cricket species we had found in basements. My idea was to use the biology of the camel crickets to predict which use they might have for humans.

Camel crickets and other basement-dwellers, including silverfish, have moved into our homes with their adaptations to cave life intact. They are able to feed on organic matter that seems as though it should be inedible. Silverfish in basements, for example, have been noted to eat plant tissue, sand grains, pollen, bacteria, fungal spores, animal hairs, skin, paper, rayon, and cotton fibers—call it civilization's smorgasbord. The diets of basement-dwelling camel crickets are likely to be similar.[1] Not only are such foods relatively devoid of nitrogen and phosphorus, which is true of the food available in quite a few ecosystems, they are often also lacking in easy-to-digest carbon. Plants and photosynthetic microbes fix carbon from the air; that carbon then forms the base of the food web of most ecosystems. But caves and basements lack light, which means little carbon is fixed, so there is little carbon to be found (except where bats are present and pooping, which sometimes happens in caves but hopefully isn't going on in your basement). In the context of the dearth of easy-to-digest carbon and other nutrients, cave animals have evolved bodies that require fewer nutrients. Cave life has favored, again and again, the evolution of animals without eyes (eyes take lots of energy to build), without pigment (pigment is often expensive), and with light and porous bones (if they have them) or thin exoskeletons (if they don't). As I thought about the potential usefulness of camel crickets, I had an idea. What if cave animals, including camel crickets, silverfish, and other species, have, in addition to all of these losses, also gained something, namely, ways of getting every last bit of energy out of the food they do find? They might, for example, rely on special gut bacteria able to break down compounds in their food that their own digestive enzymes can't handle.

If camel crickets have special bacteria in their guts for breaking down hard-to-digest compounds, we might be able to find industrial uses for those same bacteria. Maybe we could find useful bacteria in camel cricket guts, figure out how to grow them in the lab, and then find companies who would be interested in growing those bacteria to help them get rid of hard-to-break-down waste, such as plastic, or even to turn such waste into energy. It was a longshot. But what the heck, I had tenure.

To test the idea, we needed to take a census of the bacteria found in basement insects. Three groups of bacteria could be revealed in such a study. Some of the bacteria in the guts of insects or on their exoskeletons are picked up by chance and don't necessarily help the insect in any way, but nonetheless move with it from place to place. When a house fly lands on any surface, its sticky leg hairs inevitably become covered in bacteria. When it feeds, its gut fills with bacteria. These accidental passengers are then redistributed to each new place the fly lands, each place its feet touch, each place it poops, each place it regurgitates.[2] But we didn't want to study the bacterial species on the outsides of house flies, the ones just catching a ride.

A second group of specialized insect-dependent bacteria has evolved long-standing relationships with insect hosts so intimate that, in many cases, these bacteria no longer possess the ability to live without the insects.[3] Their genomes are reduced, ratcheted down to only the genes most necessary to their host insects, almost as if the microbes have become part of the insect. *Camponotus* ants depend on *Blochmannia* bacteria for vitamins not available in their food.[4] But for our purposes, although the bacteria inside the cells of insect hosts—be they weevils, flies, or ants—are fascinating, they wouldn't be useful to industry because they would be nearly impossible to grow and work with.

We wanted to focus on a third group of bacteria. We'd consider bacteria specialized to some extent for life with the insects but still capable of life on their own (in Petri dishes in our lab, for instance, or in industrial vats). Within this group, we'd focus on the subset of species that were able to independently break down hard-to-degrade carbon compounds. These would be bacteria we expected to be potentially common in the insects but rare elsewhere in the world, bacteria that might have been missed by other researchers, bacteria that were neither too widespread nor too rare—the ones that were, like the warm porridge, just right.

All we had to do now was grow the bacteria in camel cricket guts on hard-to-break-down compounds. Humans produce many industrial compounds that are relatively hard to break down. In some cases, these compounds (including plastics) are intentionally hard to degrade. This becomes a problem when we dispose of such products, and it is the reason for the great island-size rafts of plastics now floating throughout the seas. In other cases, the long-lived compounds are simply industrial by-products. Breaking down any of these pollutants would make the camel crickets very good for something.

As a biologist trained in basic (not applied) ecology and evolution, I needed help figuring out where to start with this new project. I emailed Amy Grunden, who works in the building next to mine in the Department of Plant and Microbial Biology. Amy's work focuses on the use of microbes found in nature to meet industrial challenges. For example, she has worked on the industrial application of microbes from deep-sea vents to detoxify contaminants from pesticide applications and chemical warfare agents.[5] When I asked her if she had any ideas for our experiment, Amy said, "Sure, why don't we see if any of the bacteria from the camel cricket can break down black liquor?" at which point I had to, when no one was looking, google "black liquor."

Black liquor is the toxic, black-colored liquid waste product of the paper industry. It is what is left over after you turn a tree into a white sheet ready to be inserted into your printer. It is composed of lignin, a messy carbon compound responsible for making wood hard (lignin also prevents your house from rotting away as soon as it is built), in a mix of soaps and solvents. Because of these soaps and solvents, black liquor is as alkaline as lye (it has a pH of around 12). Because black liquor is toxic and cannot legally be released into the environment in the United States, paper mills burn it, which is what makes paper mills smell like rotten eggs. Amy thought finding bacteria to break down black liquor would be useful, so we set out to find some. Stephanie Mathews, then a graduate student in Amy's lab (later a postdoc who worked with both of us and now an assistant professor at Campbell University), got to work testing some of the samples from camel crickets and from the larvae of a hide beetle species (*Dermestes maculatus*) that feeds on carrion but that is also known to be able to eat harder-to-digest foods. Stephanie and MJ Epps worked together. MJ knew the insects. Stephanie knew the bacteria. Everything was perfect—except for certain biological realities.

What Amy did not tell me when we started on this endeavor was how very difficult it was likely to be to find bacteria able to break down the lignin in black liquor. Only a handful of species of bacteria—approximately six—out of the ten million so far known to exist had been shown to break down lignin.

Fungi can degrade lignin into easier-to-use, smaller carbon compounds. Scientists call the degradation of lignin by fungi "white rot," and the fungi that can do it are "white rot fungi." The decomposition of wood in forests depends on such fungi. If it weren't for these fungi, old trees would never break down. Yet, whereas white rot fungi are very useful in nature, they are tricky to use industrially. They make mushrooms, they develop webs of hyphae, they grow very slowly, *and* they are a real mess, so everyone who has tried to use them to break down lignin, either to produce energy or to get rid of waste such as black liquor, has eventually given up. Bacteria would be easier to work with, but all six of the bacterial species that can break down lignin have proven, for one or another reason, challenging too. And no one (with the exception, it would turn out, of Stephanie during her thesis work),[6] anywhere, had ever found a bacterial species or a fungal species able to break down the lignin present in black liquor.

As Stephanie and MJ got to work, I hoped for a great discovery. But if I'd paused very long to think about the odds of success, I would have realized they were very low. But I didn't really pause and so it never dawned on me just how long a shot our plan was. MJ didn't know either. And Stephanie is endlessly hopeful, so we tried.

Stephanie works quickly. In a few months, she had results. She had grown the bacteria from the insect guts on a series of compounds. She mixed the key foods for the bacteria into agar on Petri dishes, just like the ones high school science classes use. The first set of dishes contained cellulose. A second set of dishes included lignin, but not cellulose. The other dishes had other forms of microbial food. Each of the dishes had been inoculated with a drop—an aliquot—of either slurried camel cricket body or slurried hide beetle body.

Stephanie presented the results from the Petri dishes to us. Many of the species of bacteria grew on the dishes that contained cellulose as a food source and so were able to break down cellulose. Cellulose is the material out of which paper is made, and drywall, and corn stalks. Cellulose is both waste and the material we use as a key source of biofuel. That the bacteria could break down the cellulose meant that they had the potential to be useful in converting waste cellulose into biofuel, the ability to turn corn cobs and toilet paper alike into energy. Other organisms can do this trick, and some are already being used industrially, but these bacteria might be faster or more efficient than those currently in use. This was exciting, not totally unexpected and yet still cool.

On the basis of the biology of the camel crickets we'd found in houses,[7] I expected that at least some of the bacterial species in their guts might also be able to degrade lignin. At the time I was still unaware of the history of failed attempts to find lignin-degrading bacteria. Those who ignore history are sometimes doomed to repeat it. History suggested we were unlikely to succeed in finding lignin-degrading bacteria. Yet one of the bacterial strains from the camel crickets could break down lignin; in fact, the bacterial strain could live when fed nothing but lignin as a food source. So too could five strains (of two species) from the hide beetles. It was only much later that I would realize the full significance of this discovery. In one camel cricket and one hide beetle we had, collectively, found a diversity of lignin-degrading bacteria that had nearly doubled the number of strains and increased by 30 percent the number of species known to be able to break down lignin, perhaps the most common biological compound in the world. At least two of these bacterial species, including the species we would come to focus on, a relative of *Cedecea lapagei*, appeared to be new to science. To recap, we found an unnoticed, large, introduced camel cricket species in basements across North America and, in that camel cricket, we found what appear to be new species of bacteria, species able to break down lignin.

Stephanie had also tried to grow the bacteria while feeding them lignin immersed in alkaline liquid. Imagine eating wood chips while submerged in a bath of lye, and you have the idea. The dinner is unpalatable and then your skin falls off. The bath was so alkaline that it would degrade most bacteria. Nothing should have lived, much less grown. Yet, something did. Stephanie found species that could grow even in such hostile conditions. We had done the nearly impossible on the first try! It was outstanding news. In fact, all of the species able to break down lignin, including *Cedecea lapagei*, were able to break down lignin in an alkaline bath. *Cedecea lapagei* could degrade the lignin and cellulose in black liquor and turn this waste into more bacteria, which could then be turned into energy.

By understanding the biology of camel crickets in houses, we found bacteria that give every indication of being able to turn industrial waste into energy. So too with the hide beetle. The odds of finding one new species able to break down black liquor were very low, extraordinarily low, perhaps one in a hundred thousand if not one in a million. The odds of finding three species able to do that were far lower. But such calculations suggest luck was the only contributor to our success. We did get a little lucky, but we also used our understanding of the basic biology of the camel cricket to predict where we might find useful species, and it paid off. Knowledge of natural history paid off; knowledge of ecology paid off. Knowledge of the predictable evolutionary tendencies of cave organisms paid off.

Amy, Stephanie, and I continue to study how best to grow these bacteria in quantities large enough to be useful to industry. With other colleagues, we have isolated the compounds that one of these bacteria, *Cedecea*, secretes from its cells to break down lignin. We have even found the genes that the bacteria use to produce those enzymes. We are on the road to putting those genes in bacteria used often in the lab to get such bacteria to break down large quantities of lignin in a controlled way (though not very far down that road, at least yet). Stay tuned. We find ourselves in an exciting moment. The answer, then, to what good are the species in our houses is that we don't know until we study them.

After the discovery of lignin-degrading bacteria in the guts of camel crickets and hide beetles, I felt like we had provided a pretty clear answer to the question of what good the camel cricket is. This didn't mean the camel cricket or hide beetle in any particular basement offered any more value than it had before, but it did mean that collectively these species had the potential to supply benefits to society, benefits that accrue only if the species continue to exist and if we study them. But when I gave talks about this work, people wondered if we hadn't just, by chance, picked the two arthropod species out of thousands in homes with some use to humanity. We'd grabbed at the low-hanging fruit. The only way to know for sure was to search other arthropod species for their uses too, so that is just what we did. We've set out to systematically consider the better-studied species in homes and what uses they might have.

The most obvious next step would have been to continue to search insects for bacteria able to degrade industrial waste. Book lice, for example, seem likely to have totally novel enzymes able to break down cellulose, enzymes that would help the biofuel industry. That would be easy to check.[8] Similarly, drain flies, whose larvae live in drains, are able to live on food waste in an extreme environment (the drain), one that is wet, and then dry, and then wet again. Recently, a study found that silverfish and bristletails, ancient insects also associated with caves that we found to be common in homes, have unique enzymes in their bodies able to break down cellulose.[9] We could study the silverfish and bristletails. Or we could look to other beetle species. We found two useful bacterial species in an individual hide beetle of the species *Dermestes maculatus*. One could search this beetle species more exhaustively. Or one could search its relatives, the other species of dermestid beetles. More than a dozen species of dermestid beetles live in Raleigh homes alone. Each of these likely has unique microbes. No one has checked any of these beetles for their unique microbes! I guarantee that some of these beetle species in homes have bacterial species in their guts with the potential to transform one or another industry. Doing so would be enough for an entire (and very interesting) career.

But the truth was, having identified one kind of use, one value that arthropods in homes can have, I wanted to explore other, totally different kinds of uses. Doing so blindly would have been ridiculous. But we were no longer blind. We'd learned three lessons.

**Figure 8.1** Drain flies are one of the species commonly found in homes that has received very little attention from scientists. Drain fly adults are beautiful. Drain fly larvae are not beautiful, but they are likely to host microbes able to degrade cellulose or even lignin. *(Modified from original photograph by Matthew A. Bertone.)*

The first lesson was to not assume someone else had studied the species around us, however common they might be. The second lesson was that to identify the use a species might have, it was key to know enough about its biology to surmise what abilities it might have. This meant that most of the species in houses, much less out in the wild world, could not yet be searched for their uses because we don't know what most arthropod species eat, much less any other details about their biology. The third lesson, one I'm eager to teach my own students, was that if ecologists and evolutionary biologists don't help figure out the uses these species might have, no one else will either. Let's call that third lesson a hypothesis, yet it is a hypothesis backed by a career of work with ecologists.

While walking to work, I found myself looking at each species on my path and pondering what use it might inspire. My students, postdocs, and collaborators pondered too. Together, for example, we asked whether we could figure out ways to make new cutting devices and brushes inspired by the cutting devices and brushes arthropods use. Grain beetles, for instance, have mandibles capable of breaking through seed coats that would seem impossibly tough given their size. In part, grain beetles manage this because they have metal-reinforced mandibles; the mandibles are ideally suited to cutting.[10] The shape and composition of the mandibles offer insights into how we might design new cutting tools—an entire line of cutting tools inspired by the mandibles of insects. Or a line of brushes. Most arthropod species have brushes on their legs and elsewhere to clean their eyes and other body parts.[11] One could also imagine using insect brushes as inspiration for brushes to be used in industrial production lines or even just to be used on human hair. It would be cool to use a hair brush inspired by the brush on an ant's leg. Or it would be cool if I had hair.

We also began to search the arthropods in homes for new kinds of antibiotics. Humans aren't discovering antibiotics fast enough to keep up with the rate at which bacteria are evolving resistance to the antibiotics we have. Maybe we could enlist arthropods in the search for new antibiotics, arthropods such as house flies. The mother house fly deposits bacteria such as the species *Klebsiella oxytoca* with her eggs. The *Klebsiella* bacteria produce compounds that kill fungi and help hungry young flies outcompete those same fungi for food. Such bacteria seem likely to be producing antibiotics of value to humans in the control of fungi, but they have never been studied in this context.[12] They could be. But, in terms of finding new antibiotics, house flies are just the beginning. Many ants produce antibiotics out of their venom glands, the glands just above their first set of shoulders. Decades ago a series of studies focused on isolating these antibiotics from a few ant species of giant bulldog ants (*Myrmecia* spp.) found in Australia;[13] the compounds from the bulldog ants showed promise clinically as new antibiotics for use in human medicine. When I was a graduate student, I wanted to follow up on this work. I didn't because I assumed someone else would and all the work would be done. Fifteen years have passed. The work is not done. Working with Adrian Smith at the North Carolina Museum of Natural Sciences and Clint Penick at Arizona State University, among others, we have begun to survey the ants of Raleigh, North Carolina, to figure out which species produce antibiotics. We began with the prediction that species that make large colonies or that live in the soil (where they might encounter many pathogens) would be the most likely to produce strong antibiotics. This wasn't the case. Instead, the species with the most effective antibiotics tended to be of one genus, *Solenopsis*, the genus to which fire ants belong but so too the species *Solenopsis molesta* (the thief ant). Thief ants are very common in kitchens. We found that thief ants produce antibiotics effective against bacteria closely related to the methicillin-resistant *Staphylococcus aureus* and its relatives.[14] This is to say that one day the ant in your kitchen may well save someone you know and love from a deadly skin infection.

Meanwhile, recent studies have revealed that the physical structure of the bodies of some of the insects common in backyards, if not houses per se, can either disfavor or favor specific species of bacteria. Both cicadas and dragonflies have tiny knives on their wings that dice bacteria to bits. These structures are now being replicated on building materials with the idea of making those materials antimicrobial in a way that bacteria cannot evolve to resist (it is hard to evolve resistance to a tiny knife). We wondered whether we could try the opposite, namely, to study arthropods to find ways of making surfaces that favor and house beneficial bacteria. Many ants seem to do just that on their exoskeletons. Inspired by these ants, we imagined fashioning a probiotic dress. We've made some headway, but we aren't there quite yet. In the end, we are just a dozen or so people in my lab—a few more if you include our friends—we can only do so much. Imagine large teams of people dedicated to finding the uses of the species around us every day. Imagine an institute with no other goal; of this I dream.

**Figure 8.2** An American grass spider (*Agelenopsis* sp.), one of the common spiders found around houses in North America and harmless to humans, on the threshold of a home in Raleigh, North Carolina. *(Photograph by Matthew A. Bertone.)*

MANY OF THE greatest values of the arthropods in our homes seem to be associated with species people dislike or even fear, species such as spiders and wasps. These arthropods provide valuable ecosystem services in and around houses: spiders eat pests, as do wasps. Wasps also serve as pollinators. But wasps and spiders are also terrific targets for new industrial applications. Already, spider silk has spawned efforts to produce similar materials commercially for human needs as well as efforts to build buildings the way that spiders build their egg cases and other structures, in layers and from the inside. It isn't just spider silk that is inspiration; the spigots spiders use to spray silk have the potential to inspire new approaches to three-dimensional printing. House spiders were 3-D printing long before it was cool. I suspect that if one locked a dozen spider biologists and a dozen engineers and architects in a room for a week (and maybe some spiders too), the result would be many innovations.

For my lab, wasps have proven to be a vast fount of discovery. Just as with camel crickets, we started to study the potential value of wasps in response to someone's question. In October of 2013, Jonathan Frederick, organizer of the North Carolina Science Festival, asked if we could find a new yeast that could be used to brew beer for the festival. Gregor Yanega, an expert in the biomechanics of hummingbird beaks and a postdoc in my lab at the time, suggested we turn to wasps. Gregor based this guess on two things: his knowledge of wasp biology and a recent paper that demonstrated that in vineyards wasps carry yeasts to grapes.[15] The yeasts in vineyards overwinter in the guts of the wasps, and then, once the grapes are on the vine, the wasps, quite unintentionally, carry yeasts from grape to grape. The yeasts help start the fermentation process when the grapes are harvested. It now appears that the original habitat of beer and wine yeasts, long before humans started to produce either beverage, was the guts and bodies of wasps, wasps like those that can still be found nesting on homes and other buildings around vineyards. We borrowed the yeasts from them. Gregor thought we could borrow a few more.

Searching wasps for more yeasts, those that the first farmers might have missed, was a good idea. But it was hard to implement. Who was going to go gather wasps, much less find their yeasts? Fortunately, at the time, Anne Madden had just joined the lab. Anne had spent years studying wasps. As a PhD student, Anne spent many hours hanging upside down in barns and on ladders next to eaves so that she could cut down live, buzzing wasp nests, which she would then drop (quickly) into a bag, throw on her back, and take back to her lab on her motorcycle. Anne had also, in a former life, spent years studying yeasts, particularly those of use to industry. If anyone would be able to find new yeasts in wasp nests, it was Anne.

Anne checked wasps for new yeasts. She found them. On wasps and their relatives, Anne has found more than a hundred kinds of yeasts. One was from a wasp nesting on Anne's porch at her apartment in Boston. These yeasts have amazing abilities. One can produce sour beer in a month rather than the years it often takes to make such beer.[16] You can now buy beer made with that yeast. Thanks to Anne's work, related yeasts, found in different wasps, appear to be good at making breads with new aromas and flavors. Anne thinks one possible reason why searching wasps for yeasts works so well is that the wasps themselves are using the odors produced by yeasts to find sugar sources.[17] The wasps sniff out yeasts to find sweetness. We search out wasps to find yeasts. It is a good relationship, one we hope to build on in the coming years.

ULTIMATELY, FINDING USES for the species in our homes is relatively easy. Actually documenting those uses well and bringing them to market is harder, but not impossible. The technical barriers can all be overcome with patience and money. This raises the question of why much more hasn't already been done. Why don't we have a catalog of the values of each of the species we wake up to every day? I think there are three reasons.

One reason, as I noted in the previous chapter, is that we seem to be blind to the species closest to us, and so don't even notice them. We have to see the species to study them and know what their uses might be. The second reason is that although ecologists and evolutionary biologists have spent a century talking about the "potential economic value" of species, they haven't made the effort to find those values. They assume someone else will do the work. Ecologists tend to appreciate species for aesthetic reasons, or, more simply, because they exist. With this mind-set, the uses species might have are beside the point. So, whereas my friends who work in industry think it quaint that I study insects, my friends who work on the ecology of insects think it quaint (or worse) that I work with industry. It is hard to do work that your friends don't value. That neither ecologists nor applied biologists value work at the interface of ecology and industry creates the third reason we haven't yet cataloged the potential of nearby species, namely, that most efforts made to find the uses of species have searched species randomly, one by one. This is a mistake. It has proven, in some cases, to be an extraordinarily expensive mistake. Millions of dollars have been spent searching Costa Rican rain forests, species by species, for new cancer drugs. This is the wrong way to do it. We need to let biology guide our searches. We need to use everything we know about the ecology and evolution of species to predict which are the most likely species to have particular uses. If we can finally get around these barriers, I think it is possible to combine ecology and knowledge of evolution to vastly speed up the systematic search for the uses of species, increasing our ability to rely on nature's innovations, and, having done so, perhaps, to further value the species we see around us every day. If someone asks me what good a camel cricket is now, or a wasp, or even a mosquito, I pause to think about the biology of the species in question. I pause, consider, and hypothesize, and then, once I get back to the lab, I get to work.

Of course, for any of this to work we need to know about the biology of the species around us, so we need to get to studying thousands of species of arthropods in homes (as well as the tens or even hundreds of thousands of smaller species). Inasmuch as humans now live nearly everywhere, studying the life in homes well would be a major step forward in understanding life in general. We have work ahead of us, though. I suspect that fewer than fifty arthropod species in homes (to say nothing of the bacteria, protists, archaea, and fungi) are well enough studied for us to even guess at what use they might have. So when you see the insects flying around your home, pay attention, and instead of asking "What use does this species have?" ask, "What use might I find for this species?" The burden is on us, not nature, to make the most of what evolution offers us. The burden is on us, too, to save the species around us so that when we figure out their use they are still around to help us.

Even after thinking about the values of the species in your home, even after realizing that beer and wine exist thanks to insects, if the first thought you have upon hearing about all the kinds of arthropods in houses is about how to kill them, you aren't alone. King Tut was buried with a fly swatter, his subjects apparently sure that whatever the afterlife might hold, whatever luxuries and pleasures, it would also inevitably have house flies.[18] Living ancient Egyptians also used fly swatters and plants as pesticides.[19] Cultures around the world have found ways to fight back the arthropods in their homes. Important battles have been won, particularly against some of the species that actually do cause serious problems. Garbage pickup and the piping of sewage away from homes have led to decreases in the abundance of waste-loving species able to carry disease. Mosquito nets help exclude species that vector malaria and in doing so save lives. Yet, the broader war has proven uneven and full of unintended consequences, in no small part because the species humans try hardest to kill have proven able to very rapidly evolve.

# 9

# THE PROBLEM WITH COCKROACHES IS US

You must not fight too often with one enemy or you will teach him all your art of war.

—NAPOLEON BONAPARTE

I am almost convinced (quite contrary to the opinion I started with) that species are not (it is like confessing a murder) immutable.

—CHARLES DARWIN

YOU CAN LEARN to be interested in the insect life around you and realize most of the arthropod species nearby are interesting, poorly studied, and more likely to help control pests than to be them. Or you can go to war. The modern way to wage such war is with chemistry. But be warned: if you decide on a chemical war, the battles are not evenly pitched. Not even close. To each round of new chemicals we apply, the insects we attack respond by evolving via natural selection. The more aggressive the attack, the faster the evolution. Insects evolve faster than we can understand how they have evolved, much less counter. It happens again and again, especially among those pests we try hardest to kill, pests such as the German cockroach (*Blattella germanica*).

The pesticide chlordane was first used in homes in 1948. It was a wonder pesticide, one so toxic to insects that it was thought to be invincible. By 1951, however, German cockroaches in Corpus Christi, Texas, were resistant to chlordane. In fact, the roaches were a hundred times more resistant to the pesticide than were laboratory strains.[1] By 1966, some German cockroaches had evolved resistance to malathion, diazinon, and fenthion. Soon thereafter, German cockroaches were discovered that were fully resistant to DDT. Each time a new pesticide was invented, it was just a few years, or sometimes just a few months, before some population of German cockroaches evolved resistance. Sometimes, resistance to an old pesticide would even confer resistance to a new pesticide. In those cases, the battle was over before it even started.[2] Once they evolved, resistant strains of cockroaches would spread and, so long as the pesticide continued to be used, thrive.[3]

Each of these tit-for-tat responses of cockroaches to our sinister chemical ingenuity was impressive. Lineages of cockroaches were rapidly evolving entirely new ways to avoid, process, or even take advantage of our poisons. But these responses were nothing compared to what was recently discovered in the building next to my office. The story of that discovery began more than twenty years earlier on the other side of the country, in California, where it involved two main protagonists: an entomologist named Jules Silverman and a family of German cockroaches named "T164."

Jules was tasked with studying German cockroaches. He was working at the Clorox Company Technical Center in Pleasanton, California.[4] The place was like any other science-based industry, except that what rolled off of the assembly line wasn't chocolates but devices and chemicals for killing animals. Jules focused on killing cockroaches, especially German cockroaches. German cockroaches are just one of many species of cockroaches that have moved into homes to live alongside humans. As one cockroach specialist reeled off to me at a meeting, "You've got your American cockroaches, Oriental cockroaches, Japanese cockroaches, smoky-brown cockroaches, brown cockroaches, Australian cockroaches, brown-banded cockroaches, and, well, quite a few more."[5] Most of the thousands of cockroach species on Earth do not and cannot thrive in homes,[6] but this dirty dozen seems to have abilities that predispose them to thrive indoors. Several of these species, for instance, can reproduce parthenogenetically:[7] a female cockroach can produce female offspring without any help from a male.[8] Though each of the cockroach species that can be found indoors has some adaptation aiding it in life with humans, the German cockroach has the complete package.

When a German cockroach finds itself in the wild, it is a weakling. It is eaten. It dies of hunger. Its young are blighted, afflicted, and unsuccessful. As a result, there are no wild populations of German cockroaches—not anywhere. The German cockroach is only strong and fecund in our presence, indoors. Perhaps this is one of the reasons we have come to dislike them so very much. They like the conditions—warm, not too dry, not too wet—we like. They like the foods we like.[9] They can, like us, even suffer from loneliness.[10] Whatever the reason we don't like them, we really don't have that much to fear from them. German cockroaches can carry pathogens, it is true, but not any more so than your neighbors or children carry them. Also, no cases have yet been documented in which someone has actually gotten sick from a pathogen spread by a cockroach, whereas people get sick every minute of every day from pathogens spread by other humans. The most serious problem posed by German cockroaches is that they are, in great densities, a source of allergens. In response to this real problem, and the many perceived problems, we have spent enormous resources trying to kill German cockroaches.

Just when our battles with German cockroaches began is hard to say because the dead bodies of cockroaches do not hold up well in archaeological sites (at least when compared to those of beetles). Also, people tend to study how to kill German cockroaches rather than the other details of their biology. The closest known relatives of German cockroaches are two species of Asian cockroaches, both of which live primarily outdoors. These two species fly well, feed on leaf litter and other insects, and, in some places, are regarded by farmers and scientists as beneficial agricultural insects.[11] Originally, the German cockroach was probably like these wild cockroaches. Then it moved in with humans.[12] When it did, it abandoned flying, began to reproduce more rapidly and live more gregariously, and adapted in other ways so as to be most successful in those conditions most preferred by humans. Then it spread.

The German cockroach appears to have made its way through Europe during the Seven Years War (1756–1763), a time when people were traversing Europe with containers large enough to hold quite a few cockroaches. Just who moved the German cockroach around is unknown.[13] The father of modern taxonomy, Carl Linnaeus, asserted it was the Germans. Linnaeus was Swedish. The Swedes fought against the Germanic Prussians, so Linnaeus thought that "German cockroach" seemed a fitting moniker for a species that even he didn't like.[14] By 1854, the German cockroach was in New York City. It now lives from Alaska to Antarctica, having moved with peoples of nearly every nation and their boats, cars, and planes.[15] It is surprising German cockroaches aren't yet on the space station.

In places where the temperature and humidity of homes and transport vehicles still vary with the seasons, the German cockroach coexists in homes with other cockroach species,[16] some of which (such as the American cockroach) may have associated with our lineage since the days we lived in caves.[17] But in dwellings in which humans have invested in central cooling and heating, the German cockroach comes to dominate. Other cockroaches, in response, tend to become rare. Until recently, for example, the German cockroach was uncommon in much of China, but as China began to heat its transport trucks in the north (where it is cold), the conditions in those trucks became warm enough for the German cockroach and it moved north. When China began to cool its trucks in the south (where it is hot), the trucks became cool enough for the German cockroach and it moved south. Having arrived, the German cockroach now finds enough heated apartments in the north and cooled apartments in the south to thrive. Throughout China, and around much of the rest of the globe, as more apartment and house dwellers have invested in central cooling and heating, the German cockroach has become ever more widespread and abundant.[18]

By the time Jules Silverman started work at the Clorox Company twenty-five years ago, populations of the German cockroach were already on the rise. Jules's job was to develop new chemicals to kill German cockroaches. The best thing on the market at the time was roach bait. You know roach baits. They are little sugar treats for cockroaches laced with pesticides. Cockroach baits enable us to poison cockroaches without having to spray poison around a house. In theory, the baits can be made with any of the sugars to which roaches are attracted: fructose, glucose, maltose, sucrose, or maltotriose. In practice, glucose is always used in the United States. It is cheap, and the attraction of cockroaches to glucose is strong. The German cockroaches living in the United States are used to glucose. As much as 50 percent of their diet is composed of carbohydrates, and most of those calories come primarily from glucose. It is the same stuff we feed ourselves in huge quantities in the form of corn syrup. We bait our children to dinner with the promise of dessert using the same substance we use to bait cockroaches to death.

In those first years at Clorox, Jules realized that something had gone wrong in one of the apartments where his friend, the field entomologist Don Bieman, had placed baits. It was apartment T164. In T164, the German cockroaches weren't dying when Don put baits out.[19] They lived. He put out more baits. They still lived. When tested in the lab, cockroaches from T164 died when exposed to the poison used in roach baits at the time (hydramethylnon). The poison killed them in the lab, but not in the apartment. Don told Jules that it almost seemed like the cockroaches in the apartment were repulsed by the baits. In the lab, Jules tested the attraction of the cockroaches from T164 to the different components of the baits. The first and most obvious possibility was that the cockroaches had begun to somehow avoid pesticide in the baits. But Jules's experiment showed that the cockroaches did not avoid the pesticide in the lab. Nor did they avoid the emulsifiers, binders, or preservatives in the baits. The only thing left to check was the sugar in the bait: glucose, aka corn syrup. It would be very surprising if the cockroaches avoided glucose, which would require avoiding a food to which cockroaches and most other animals have been attracted for millions of years—sugar. But that is exactly what had happened. The cockroaches were avoiding glucose. It wasn't just that they weren't attracted to it; they were repulsed. Disgusted. But they were still attracted to fructose. Perhaps, Jules thought, this particular population of German cockroaches (which would come to be called T164) had learned. Somehow, they had acquired some sort of superpower. Hell hath no fury like a clever German cockroach (except perhaps billions of clever German cockroaches).

Jules could test the idea that the cockroaches were learning. If they were, their babies—each one fleshy, pale, unprotected, and ignorant—should be attracted to the traditional baits, as should their grandbabies. The babies and grandbabies, upon birth, would not have yet had a chance to learn. Jules tested whether the babies and grandbabies were attracted to glucose. They were not. The cockroaches had not learned; they were born with an innate aversion to the sugar glucose, in particular. The only way to account for this dislike of glucose was to hypothesize that this aversion was genetic and had evolved. Jules performed simple genetic experiments to see how the aversion to glucose was inherited. He bred cockroaches averse to glucose with those still fond of it, then crossed their offspring with the parent that preferred glucose. Those crosses suggested that the gene or genes that controlled the aversion to glucose were dominant, though incompletely so.

**Figure 9.1** German cockroaches from Jules Silverman's T164 colony feed on a sample of peanut butter (with no sugar added) and carefully avoid the glucose-rich strawberry jam. *(Photograph by Lauren M. Nichols.)*

Imagine that a family of German cockroaches moves into a big apartment building. Over time, a few cockroaches can become many more. Every six weeks, a female cockroach can produce an egg capsule containing up to forty-eight eggs. At this rate (which is fast relative to human reproduction but pretty ordinary for an insect), even if an individual female German cockroach lives only long enough to produce an egg case twice she can nonetheless give rise to ten thousand descendants in a year.[20] When an exterminator places baits throughout the building and all of these thousands of cockroaches die, no evolution occurs. No particular versions of genes are favored relative to any other versions. The story is over until German cockroaches colonize the building anew and are baited again. If some of the cockroaches survive, however, and if their survival relates to a trait encoded in their genes but absent in those of the cockroaches that died, then the use of baits would actually favor the surviving cockroaches and their versions of genes. This is what Jules came to believe happened, that some version of some gene or set of genes made the T164 German cockroaches less attracted to glucose, or even repulsed by it. The T164 cockroaches had been favored, he thought, by the glucose baits and then, in their success, had rendered the baits useless.

Jules next sampled German cockroaches from around the world for glucose aversion. In many places where glucose baits had been used, from Florida to South Korea, the cockroaches had evolved aversion. And they appeared to have evolved this aversion independently in each of these places. Jules tried to repeat this finding in the lab to see whether he could actually experimentally cause evolution to occur. He gave German cockroach populations glucose baits laced with insecticide. The changes he saw in the lab resembled those occurring in the wild: over relatively few generations, glucose aversion evolved. He wrote a series of papers about his findings.[21] He patented a series of roach baits based on fructose.[22] He thought he might be about to launch the careers of many evolutionary biologists who could help him figure out the details of what seemed to be very rapid evolution occurring in German cockroaches.

But while pest control companies responded to Jules's discovery by using his newly patented fructose baits, evolutionary biologists seemed to ignore the work. Jules thought he knew why: he could not explain the mechanism by which the German cockroaches evolved to be able to avoid glucose, which kinds of genes were affected, what those genes did, or even how it all happened so fast and so repeatedly. With time, though, he thought he could figure it out, and so, for years and then decades, he maintained the descendants of the German cockroaches he first studied in case one day they might be needed. Different people have different keepsakes. One man's snow globe is another man's perpetual cockroach colony.

While Jules waited for more insight about the German cockroaches, he went on to study other pests and their evolution. He moved to North Carolina State University in 2000, where he spent the years from 2000 to 2010 studying a population of Argentine ants (*Linepithema humile*) that had spread in the southeastern United States, from yard to yard and then into one building after another. He also studied the odorous house ant, *Tapinoma sessile*.[23] For ten years, he didn't touch a cockroach, except to continue to feed his pet colony—the descendants of the population from apartment T164—on which his biggest and largely ignored discovery was made.

In some ways, the story of the German cockroach is unique. There is no other species quite like it. But in other ways, it is just a kind of heightened example of what is happening with many of the species in our homes. Evolution can be wonderfully creative—whimsical even—in what it produces, but it also has a kind of predictability. That predictability relates to the tendency of evolution to produce convergent forms in unrelated organisms. Wings evolved independently in insects, bats, birds, and pterosaurs. Eyes evolved once in our lineage and, independently, in the lineage of squids and octopuses. In plants, trees evolved again and again, as did spines and fruits. But so did far more unusual features, such as plant seeds with tiny fruits intended for ants. The ants carry those seeds back to their nest, eat the fruit, and discard the seeds in their garbage piles, where the seeds germinate. Such ant fruits have evolved independently more than a hundred times.[24] Key to predicting just which tricks evolution will repeat is an understanding of the opportunities available to species and the challenges in taking advantage of those opportunities. In our houses, the opportunity that is available is the potential to eat off our bodies, our food, and our houses themselves. The challenges have to do with arriving in our homes and surviving our assaults.

Certain circumstances lead to rapid adaptation to biocides: when the species we are trying to kill are genetically diverse (or have ways of borrowing new genes from other species); when the biocide kills nearly all (but not all) of the individuals of the species we are trying to kill; when the organisms are exposed to the biocide repeatedly (or even chronically); and when the competitors, parasites, and pathogens of the species we are trying to kill are missing.

These conditions are met especially well for German cockroaches. But they are also met for nearly all of the species in our homes that we most actively try to kill and keep out. As a result, homes are one of the places in which evolution is happening most rapidly, though rarely in ways that are to our benefit.

Resistance to our pesticides has evolved among bed bugs, head lice, house flies, mosquitoes, and other common insects in houses. Natural selection can offer us great benefit, but only if we make decisions that reflect our knowledge of how it works. We don't tend to. As a result, in our daily lives, natural selection is far more likely to prove dangerous than beneficial for us, and its dangerous gains are accumulating faster than our understanding, faster too than our ability to fight them. In short, the pests are winning so many victories that those evolutionary biologists who study resistance have been busy. They have found a great deal to do in the years since Jules discovered the glucose-averse German cockroaches, a great deal to be done even without studying those German cockroaches themselves.

The trouble was that resistance was evolving again and again, and, when it did, resistant forms replaced susceptible forms and spread. When new traits evolve on remote islands, they often stay there. Vampire finches evolved just once and never spread. Komodo dragons are confined to just five islands. But if a species evolves resistance to a biocide or other control in a home, that species can readily move to any other home in which the same control measure is being used and even those in which it isn't. In rural environments, such spread of resistant species might be slow. But in cities, it can happen rapidly because apartments and houses are close together, because the movement of people, boxes, trucks, ships, and planes from place to place is frequent and rapid, and because the transport vehicles themselves are ever more similar to homes. Inasmuch as cities are the future, so too, then, is this ability to spread. Even though human social networks often fall apart in cities, with feelings of loneliness and isolation very much on the rise, the resistant pests are able to stay connected. Their movement is a kind of river, a river of our own making that flows through our windows and under our doors.[25]

Whereas resistance is quick to evolve among the species we don't like, it is less likely to evolve among the rest of life. This is doubly problematic. The first problem it poses is the simple loss of the biodiversity around us in the world, the biodiversity on which wild ecosystems depend. A recent study found that over the last thirty years the biomass of insects in Germany had declined in wild forests by 75 percent. The jury is still out on just what caused this decline, but many scientists think that pesticide use is likely to have played a role—pesticide use on agricultural fields but also in backyards and in homes. The second problem is that the species that are most likely to die as a result of pesticides are the beneficial species in general, including, for example, pollinators and species ecologists call natural enemies, the natural enemies of the pests we are trying to control.[26] The natural enemies of the pests in our homes are very often, whether you like it or not, spiders.[27] If you kill the spiders in your home (and this is precisely what we do with many kinds of pesticide applications), you do so at your own expense.

As children, we learn about the old woman who swallowed a spider after swallowing a fly. That case didn't turn out well (spoiler alert: she died). Others have turned out better. In 1959, a researcher in South Africa, J. J. Steyn, was trying to figure out how to control house flies in homes and other buildings. House flies (*Musca domestica*) are ancient associates of humans, having journeyed around the world with Western civilization to nearly every region in which humans live. But they can be a real problem, especially when sanitation is poor. Much more so than German cockroaches do, they vector pathogens, including many that cause diarrhea and are associated with more than five hundred thousand deaths a year. They, like German cockroaches, also rapidly evolve. By 1959, house flies in South Africa were resistant to DDT, BHC, DDD, chlordane, heptachlor, dieldrin, isodrin, prolan, dilan, lindane, malathion, parathion, diazinon, toxaphene, and pyrethrin.

The flies had become, and remain, largely invincible to chemistry. But they weren't and aren't invincible to spiders.

J. J. Steyn gained a key insight from *The Afrikaans Children's Encyclopaedia*, perhaps while reading it to his own children. The encyclopedia noted that in parts of Africa colonies of social spiders (species of the genus *Stegodyphus*) are intentionally brought into houses to control flies and other pests. The practice of bringing social spiders into homes to control flies appears to have first been used among the Tsonga and the Zulu. The Zulu even incorporated special sticks into the construction of their homes that made it easier for the spiders to build nests.[28] The colonies of these social spiders are large, often the size of a football or a soccer ball, and easily transported from house to house by humans.

Steyn wondered whether the spiders could be used in houses again, but also outside houses and in the pens of goats and chickens, where flies were both abundant and likely to transmit disease. He tried. It wasn't hard. In kitchens, the spider webs were suspended by a string attached to a nail. Once there, they controlled flies effectively. The spider webs were also introduced into hospitals. Again, they controlled flies effectively. Steyn repeated the experiment (boldly) in the animal house of the Plague Research Laboratory. In the laboratory, the fly population declined by 60 percent in three days. In the winter, the spiders slowed down and caught fewer flies, but there were also fewer flies to catch.

From his research, Steyn concluded, "In order to help protect humans against fly-borne diseases, it is suggested that colonies of the social spider be placed in public places like markets, restaurants, milk barns, public houses, hotel kitchens, as well as in abattoirs and dairies, and especially in kitchens and latrines on all possible premises. In cowsheds, they would also help to increase milk production."[29] He imagined a world of houses filled with giant balls of spiders, a world in which flies and the diseases they transmitted would become rare, a world in which one element of traditional Zulu or Tsonga knowledge of spiders might, again, be usefully applied.

Steyn wasn't the only one with such a dream. In parts of Mexico lives another social spider, *Mallos gregalis*. This spider, too, forms large colonies (with as many as tens of thousands of individual spiders). These spiders, too, were brought into homes to eat flies, in this case by the indigenous people of Mexico.[30] Just as in South Africa, this approach was part of the traditional knowledge of local people, knowledge later discovered by Western scientists. At one point, *Mallos gregalis* spiders were even introduced into France in an attempt to control house flies. On the first try, the plan failed when the scientist went on vacation and the person left in charge of the spiders was unable to keep them well fed. The idea of a web of a giant social spiders in your house may be off-putting, but remember that every house we have ever sampled—be it in Raleigh, San Francisco, Sweden, Australia, or Peru—has contained spiders. The question is not whether spiders are in your house controlling pests, it is whether you have enough spiders, of the right species, to do the job well.[31]

Spiders are not the only species that can be used in biological control in houses. Many solitary wasp species consume nothing other than one or another specific cockroach species, but they do so in a far different way than do spiders. The wasps are tiny. They do not sting. They preoccupy themselves with hunting for the egg cases of some species of cockroaches. The wasps can smell these egg cases. When they do, the mother wasps tap the egg cases to make sure they still contain live roach eggs. If they do, she then pierces the egg case with her ovipositor (egg-laying device) and lays her eggs inside. The wasp eggs hatch, devour the roaches inside the egg case, drill a hole in the egg case, and escape like young birds tipping out of the nest. In one study of homes in Texas and Louisiana, 26 percent of American cockroach egg cases were parasitized by the wasp *Aprostocetus hagenowii*, and another portion were parasitized by yet another wasp, *Evania appendigaster*.[32] We did not find *Evania* in any houses sampled in Raleigh, but *Aprostocetus hagenowii* was very common. If you find an egg case in your house with a hole in it, it has likely given rise to wasps rather than to roaches.

**Figure 9.2** Social velvet spiders, *Stegodyphus mimosarum*, feeding on a house fly. *(Photograph by Peter F. Gammelby, Aarhus University.)*

They may be in your house, flying around now, small and entirely beneficial. Several researchers have attempted to release parasitoid wasps into homes to control roaches. All of those attempts have been, in one way or another, successful (though also typically poorly documented). Nor is it just spiders and tiny wasps that can help keep our houses in order. Another research project aims to use the fungus *Beauveria bassiana* to control bed bugs. If sprayed on a surface in a home, the spores of *Beauveria* sit, waiting. When a bed bug passes, they attach to the external layer of fats on the surface of its exoskeleton. Once attached, the fungi grow through the bed bug's exoskeleton. Once inside, the fungi proliferate in the bed bug's body cavity and kill the bed bug by simultaneously clogging and poisoning its organs and starving the rest of its body of key nutrients.[33]

In our nightmares, the wasps we release to control cockroaches deposit their eggs into our bodies, and the baby wasps develop in our body cavities, eat us from the inside, and then hatch from one or another of our orifices (or make a new one). This does not happen. These wasps are small, safe, and our allies. Similarly, we imagine that the spiders in our houses might bite us. Or even consume us. They don't, either. Spiders, too, are nearly always our allies.

Each year, tens of thousands of "spider bites" are reported around the world, and the numbers seem to be increasing. Yet spiders rarely bite humans, and nearly all of these "bites" are instead actually cases of infections due to resistant *Staphylococcus* bacteria (MRSA) misdiagnosed by patients and doctors alike. If you think you have a spider bite, ask a doctor to test whether you have a case of MRSA. Those odds are much higher. One of the reasons spider bites are rare is that most spiders use their venom exclusively or nearly exclusively on prey rather than in defense. For spiders, it is nearly always easier to flee than to fight. One study even attempted to find out how many pokes it took to get each of forty-three individual black widow spiders to bite artificial fingers (made out of congealed Knox gelatin). But the spiders wouldn't bite. After one poke with an artificial finger, none of the spiders tried to bite. Nor did any of them try to bite after sixty repeated pokes. The only time in the study that a black widow bit the artificial fingers was when the fingers were used to intentionally squish the spider three times in a row. Sixty percent of the spiders squished between two artificial fingers three times in a row bit. Even then, the spiders that did bite released venom only half of the time, such that half of the bites would not have been problematic, just painful.[34] Venom is costly to spiders, and they don't want to waste it on you; they are saving it for mosquitoes and house flies.[35]

Our use of chemistry to kill species where we live, on the other hand, comes back to bite us again and again. Spraying pesticides in houses and backyards creates what ecologists call an enemy-free space for any pests resistant to those pesticides. Our goal should be the opposite: homes full of (rather than free of) the enemies of our pests. The use of cockroach baits, for example, was supposed to be a solution to this problem. The pesticides would be consumed by the pests, but not by their predators. But then the cockroaches evolved a way around even this human innovation. Just how they evolved remained a mystery until 2011. By that point, Jules Silverman had begun to shift the work he was doing in the lab. He stopped working on cockroaches and ants and was beginning to spend most of his research time on aquatic insects. He converted his lab into a series of giant tanks filled with caddisflies and algae. He started teaching a class on aquatic insects. He was sinking, waders first, into a new phase of his life. But Jules kept feeding the cockroaches, and he continued to search the literature for ways to help solve his riddle, the riddle of the resistant cockroaches. He would soon have company in this quest.

Jules works in an aging building at North Carolina State University, a building in which air conditioners and heaters dangle in windows. The air conditioners are not for the people in the rooms but instead are used to keep comfortable the insects under study by the university's entomologists, which include Jules's cockroaches. Because the insects being studied tend to be household pests, they need to be kept in conditions similar to those found in a modern house, with constant temperature and relatively constant humidity. For the insects the climate is controlled. Each entomologist keeps a different creature. In the lab of Wes Watson, a veterinary entomologist, one can find the flies that live on the eyes of cows or the beetles that wriggle through their dung. In the lab of Michael Reiskind, an expert on the ecology of mosquitoes, blood-fed female mosquitoes lift off with each shake of the walls (the walls do shake, especially when a train goes by) and then settle back down. But the lab in which the most kinds of pests can be found is that of Coby Schal, an expert on the ways in which household pests communicate with each other. In Coby's lab, bed bugs cling to blood-filled membranes and a half dozen species of cockroaches stumble, in great densities, over one another's bodies.

Like Jules's, Coby Schal's work includes the study of cockroaches, especially German cockroaches. Coby is a chemical ecologist: he sees nature as a function of the chemical signals organisms use to communicate with each other. More specifically, he is an expert in the chemistry of cockroaches and how they communicate with each other. He has discovered, among many other things, a pheromone used by wild female cockroaches to attract males. He can put this pheromone out in a field (or even hold it up in his hand) and male cockroaches come flying toward him, only to be disappointed.[36] Jules knew of Coby's work long before the two ever became colleagues. He cited one of Coby's papers in his own first paper on cockroaches. But even once they were both at the same university, the two didn't work together on cockroaches. They teamed up to study Argentine ants, and odorous house ants, but not German cockroaches. Maybe they were just both busy with other collaborations. Maybe Jules didn't think Coby's skills were quite what he needed to resolve the questions he was most interested in. For some mix of reasons, no cockroach collaboration materialized.

And then a new postdoctoral researcher from Japan, Ayako Wada-Katsumata, arrived in the department in 2009. Postdoctoral researchers very often have skills their bosses lack. They also have more time to do research and so can, through their work, build bridges where none existed before. Such was the case with Wada-Katsumata. She had the skills that bridged Coby's and Jules's work and, in doing so, ushered forth what Jules regards as one of the most important discoveries of his career.

Wada-Katsumata's special skill is measuring how the brains of insects such as cockroaches respond to compounds they taste or smell. Before coming to North Carolina State University, Wada-Katsumata considered whether sharing food triggers chemicals associated with pleasure in the brains of ants. (It does.) She also studied the sensory experiences of cockroaches during courtship and sex. During courtship, male and female German cockroaches find each other in the dark. The female cockroach produces a chemical signal that becomes airborne and drifts through a house and attracts male roaches to her. The chemical drifts out of kitchen cupboards and up from beneath cabinets. It drifts around corners and up the stairs. Even when the lights are out, the male can find the female by chasing her odor.[37] The male and female then make contact, and when they do, the male detects other chemicals produced on the female's body. In response, he offers up a nuptial gift, a package of sugars and good odors, a kind of sexual candy filled with sugar and fat. Depending on her satisfaction with this gift (which she eats regardless), she decides whether or not to mate. When Wada-Katsumata began her work on cockroaches, the composition of the nuptial gift of male roaches was known, but the response the gift triggered in the brains of female roaches was not. To figure out the answer, Wada-Katsumata wired the taste neurons of German cockroaches, neurons located on tongue-like sensilla, to a computer and then offered both females and males different kinds of gifts. In this particular experiment, she was playing the role of the male cockroach. When she did, she found that the male's gift was perceived as a tasty food by both the male and the female, but that the food stimulated the female's neurons more strongly than it did the male's. If a male cockroach grew despondent and lonely, he could eat his own "gift" and enjoy it, but not nearly as much as a female would.

In North Carolina, Wada-Katsumata would consider almost the opposite of what she was working on in Japan. Rather than study the response of German cockroaches to something they sought out, sex, she'd study the response of the T164 cockroaches to something they avoided, glucose. Jules believed and Coby, through their discussions, had also come to suspect, the T164 cockroaches had evolved a way of responding aversively to the taste of glucose. One outlandish possibility was that natural selection had favored T164 cockroaches in which glucose triggered the "bitter" neurons on sensilla rather than the "sweet" ones. Perhaps they touched the glucose with their sensilla and their brains shouted "Bitter! Walk away!" It was already known that the sweet taste receptors of ordinary German cockroaches (what scientists call "wild type" cockroaches) respond to both glucose and fructose. But was the same still true of the T164 cockroaches? Wada-Katsumata would try to figure it out. Like a cockroach mind reader, she would test what the cockroaches perceived.

The task would consume most of her working hours. Morning after morning, she ate breakfast, traveled to the lab, gathered up a cockroach, and then corralled it into a tiny cone put on so that the cockroach's head stuck out the small end of the cone and its bulbous, swollen body protruded out the other.

Once the cockroach was set up in its cone, Ayako looked through a microscope at the hair-like sensilla on its mouth. She connected one end of an electrode to a single sensillum. The other end of each electrode ran to her computer. The electrode connected to the sensillum was surrounded by a narrow tube containing water and glucose (or whatever else she might want to offer the cockroaches in this taste test). Depending on the amplitude and frequency of the impulse on her computer screen, Wada-Katsumata interpreted whether the food she had given the cockroach, be it fructose, glucose, or anything else, triggered "sweet" neurons or "bitter" neurons on the sensillum. If she saw a fast pulse on her screen, she knew the "bitter" neurons were triggered; the cockroach perceived bitter. If she saw a slightly slower and larger-amplitude pulse, she knew that the "sweet" neurons were triggered; the cockroach perceived sweet. It was an elaborate process, one that Wada-Katsumata repeated for five sensilla from each of two thousand roaches—half of them from the T164 population, half of them wild type.

The work took more than three years. Over these three years, Wada-Katsumata sat eye to eye with these cockroach heads, testing them. They looked at her. She gave them sweets. They responded to those sweets with tiny impulses that showed on her screen. She saved the resulting data in the computer. She backed up the data. She did this both for German cockroaches that were glucose-averse (Jules's T164 German cockroaches) and for normal German cockroaches that ran higgledy-piggledy toward the stuff. It took a full day to test each individual cockroach, sensillum by sensillum. The experiments required patience, persistence, and then, when those were both all gone, something a little more. And all of this because Jules and Coby, and now Wada-Katsumata, thought that the key to understanding population T164 might have to do with what happens in their brains when they taste glucose.

Wada-Katsumata's results slowly accumulated. There was no key moment. Eventually, the answer was finally so clear no more testing was necessary. The T164 cockroaches and the wild type cockroaches both perceived fructose as sweet in the same way that the cockroaches she had studied in Japan perceived each other's sex signals as sweet. Fructose triggered their sweet neurons. The wild cockroaches also perceived glucose as sweet. All of this was as expected. But—and here was the key—the cockroaches Jules had dragged with him city to city—his tether to his former life, the T164 specimens—perceived glucose as bitter.[38]

How could this be? The only possible interpretation is that the original cockroach baits in apartment T164, baited with glucose, were so deadly that most—but not all—roaches died. And some of the roaches that didn't die were those that avoided the baits entirely because they had a version of a gene or genes that led them to perceive the bait itself as bitter. This needed to happen only once. From that single event, all of the T164 German cockroaches may well have derived. Time has added nuance to this story, though ambiguities remain. For example, Wada-Katsumata has shown that not only were the surviving cockroaches averse to glucose but also, in regions where baits were originally baited with fructose rather than glucose, the roaches evolved to perceive the fructose as bitter instead. This means that the evolution of the roaches is predictable. Evolution in light of our actions is predictable. What is not yet understood is which specific versions of which specific genes were favored so as to lead the T164 cockroaches to perceive glucose as bitter.

Wada-Katsumata is back in the lab. Jules has charged her with taking care of his cherished cockroaches from apartment T164; he has passed on the care of their line. He is pondering retirement

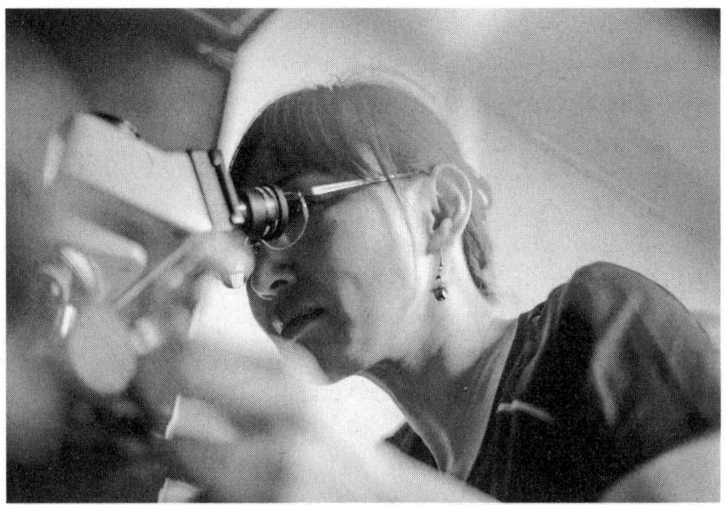

**Figure 9.3** Ayako Wada-Katsumata observing a cockroach through a microscope in the lab. *(Photograph by Lauren M. Nichols.)*

and Wada-Katsumata's career is just beginning. They would be her legacy now. With these cockroaches, she is studying how the evolution of an aversion to sugars affects the sex lives of roaches. It is an integration of the work she did before she came to North Carolina State University and her work with Coby and Jules. The long answer with all of its context and contingent details has yet to emerge because the science is slow and hard and a clear picture could well take her whole career to emerge. But the short answer is that the roaches that dislike glucose are less able to mate. Males try to attract females, but their sweet chemical telegram contains glucose, so rather than being sweet, sexually sweet, it is bitter. Consequently, the female often skips the sex and moves on. Who can blame her? Because female cockroaches are more likely to walk away from sex with bitter males, for the male cockroaches in your house a trade-off exists between being sexy and surviving. In theory, this means that when you use cockroach poisons laced with glucose, you are favoring the lineages of cockroaches less able to get it on and so less able to fertilize enormous indoor cockroach populations. In practice, though, a less-than-fully-sexy male cockroach is still sexy enough to beget millions of descendants.

It might seem that the story of the German cockroaches of population T164 sheds light only on the evolution of cockroaches themselves or on the ways in which a clever and persistent scientist can reveal what seemed to be unknowable. But just as military specialists study the battles of the past to prepare for the future, we might consider our battle with the German cockroach in contemplating our own evolutionary future.

Evolutionary biologists spend very little time writing about and predicting the distant future. It isn't that they are shy about making predictions but instead, I suspect, it is because the evolutionary future is entirely contingent on the fate of our species. Evolutionary biologists know that every species eventually goes extinct. We will too. They know that in our absence evolution will continue, as it long has.[39] It will be punctuated by the occasional disaster, as it long has been, and yet it will tend inevitably toward more diversity, more kinds of life, as has happened after every major extinction or change in the evolutionary past. In our absence, the future will unfold according to the general rules of evolutionary biology. There is horror in such a view of life, the horror of the end of our kind, but there is also a kind of solace in knowing life will go on without us and will offer up forms of existence we have yet to imagine (and won't be around to see either).

Thinking about what happens while we're still around is trickier. So much depends on the decisions we make and the innovations we offer up to the world. We now control much of the evolution occurring on Earth, albeit unwittingly and with sloppy orchestration. In light of this, the easiest outcome to consider is what will happen if we continue to make the same sorts of choices we have made over the last hundred years. These choices are, in turn, the same sorts of choices we have made over the last thousand, ten thousand, and even twenty thousand years, choices to kill what is problematic or aesthetically unpleasant and visible with ever more powerful weapons.

This is a future that is easy to imagine. It is a future in which the use of novel chemicals as weapons favors the evolution of ever more behaviorally and chemically defended pathogens and pests and leaves far behind—if they are left at all—the species that might benefit us. The pests will be resistant, but the rest of life, the rest of biodiversity, won't be. We will unknowingly trade a richness of wild species—of butterflies, bees, ants, moths, and the like—for the few resistant life-forms. The exoskeletons of those enduring life-forms will be coated with barriers that prevent toxins from entering their bodies. Their individual cells will have transporters that prevent toxins from being moved in (or special fat bodies where toxins that have moved in can be stored safely). Like the cockroaches, they might also be ascetic, forgoing the diets and maybe even the sex pheromones with which we bait them to their deaths. It is already happening, but will accelerate and become more extreme, and more global. The more homogenous and climate controlled we make our spaces, the easier we make indoor life for ourselves, and the easier indoor life will be for them too.

And whereas what evolved in the Galapagos Islands, where Charles Darwin most clearly saw the process of natural selection and its result, evolution, was animals without fear of humans, what is evolving around us is the opposite: a miniature army that knows just how to avoid us and our assaults. Indoor pests will continue to be nocturnal. They will specialize in whatever hours we do not occupy, the hours in which we fail to pay attention (we kill pests when we notice them). To some extent, this has already happened. Bed bugs evolved from bat bugs sometime when humans lived in caves. Bat bugs are diurnal; they eat at bats when bats are sleeping. Bed bugs, on the other hand, have evolved to be nocturnal, so they can eat at us when we are sleeping. So too have many cockroaches and rats turned nocturnal. Animals will also evolve to sneak through ever smaller cracks. The more we seal up our buildings, the smaller these organisms will get. The most obvious future is one in which the thousands of species of animals we now find in homes, each with an interesting story and most with no negative effect on humans whatsoever, will be gone and in their place we will be surrounded by the consequences of our actions, thousands of tiny, resistant, evasive German cockroaches, bed bugs, lice, house flies, and fleas. We will be surrounded by their diminutive army, an army that skitters away from us on its many, many legs when we turn on the lights and then, as soon as we disappear or turn the lights back off, regroups and reclaims.

10

# LOOK WHAT THE CAT DRAGGED IN

I cannot make you understand. I cannot make anyone understand what is happening inside me. I cannot even explain it to myself.

—FRANZ KAFKA, *The Metamorphosis*

And in whatever houses a cat has died by a natural death, all those who dwell in this house shave their eyebrow.

—HERODOTUS

TO THE EXTENT that we manage the animal life in our houses, we tend to try to get rid of it, as with the case of German cockroaches. But there is one exception, one very important exception—our pets. Our pets are good. They keep us happy and healthy. In exchange, we feed them. We pet them. We walk them more than we walk our own human children. In a biological world full of ambiguities, our pets are unambiguous, unambiguously good. Or at least they seem to be until we begin to consider the species that ride into our houses with our pets. When we do, everything suddenly (once more) gets complex.

When most people think of pets, they are reminded of their own domestic animals. Maybe their first pet or a pet alongside which they weathered some storm. But, as an ecologist, when I think about pets, I am reminded of the first job I ever had in science, studying beetles. I was an eighteen-year-old undergraduate student. I applied for an internship to watch monkeys. I didn't get it, so I applied for a second internship, to watch beetles. I got it. So it was that I began to help a graduate student at the University of Kansas, Jim Danoff-Burg, with his thesis.[1] Jim was studying a group of beetles that live with ant species of the genus *Liometopum;* these ants give off an odor of citrus, apricots, and slightly sweet blue cheese when alarmed (which they always are when ant biologists are poking at them). These ants build large underground nests in the desert and can be found by turning stones or searching at the base of juniper or pinyon bushes. You can find them at night without a flashlight, by smell, if you don't mind also finding rattlesnakes.

The beetles that live with *Liometopum* ants are, for all practical purposes, their pets. The beetles have evolved the ability to solicit food and shelter from the ants. Ants produce specialized compounds that they use to appease their colony mates; for example, these compounds can help calm the colony after some danger has passed. The beetles produce chemicals that mimic those produced by the ants themselves, chemicals that appease the ants, much as humans are appeased (and pleased) when petting a dog. The beetles also rub on the ants, like a cat rubbing your leg or a dog pushing against you to be petted. In doing so, they pick up the odor of the ants on their body and begin to smell like ants. Smelling like an ant is key; it keeps the ants from eating the beetles. Ants kill and eat nearly anything that moves and doesn't smell like a close relative (distant relatives, on the other hand, inasmuch as they tend to be from neighboring colonies, are eaten without qualms). Having calmed the ants and made themselves invisible, the beetles walk around eating bits and pieces of the ants' unattended food. Some species of these beetles even convince their host ants to feed them. They sit in front of the ants with their front "paws" in the air and beg.

At least part of the effect that the beetles have on the ants may be negative in that they extract a tithe of food. Though, like dogs or cats in early human societies, this food may be the leftovers the ants don't want. The beetles may also feed on pests and pathogens that live in the ants' garbage pile, which is beneficial to the ants. When working on these ants, Jim and I decided to test whether the beetles were, on average, costly or beneficial to the ants.[2] We tried an experiment in which we put ants into film canisters with and without beetles and then tallied how long the ants in each treatment survived. The difficulty with the experiment was that we had to do it while driving around in Jim's car (as we traveled from place to place looking for more sites with ants and beetles). The ants with the beetles seemed to survive longer than those without. We hypothesized that perhaps the beetles calmed the ants and prevented them from wasting energy on wild panic. That the ants might panic was understandable. After all, they were riding in a film canister through the desert in an old Toyota Tercel, bathed in the odors of their desperation and our peanut butter. The experiment suggested the potential benefit of the beetles to the ants, at least under some conditions.

The experiment with the ants and the beetles wasn't easy, but it was nonetheless easier than the same experiment with humans and pets. No one would give you permission (anymore) to put a human and a dog in a giant jar and wait to see whether humans with dogs in their jar live longer than those without dogs. Actually calculating whether our dogs and cats (or for that matter indoor pigs, ferrets, and even companion turkeys) benefit us in terms of health and well-being is difficult. Dogs with special roles, such as performing services for people with disabilities or detecting cancer, have obvious direct benefits for humans. But what about the average dog or cat, the household pet? A small number of studies have found that having a dog, and to a lesser extent a cat, can reduce stress, anxiety, and feelings of loneliness, an effect similar to the one we thought the beetles had on the ants. It is an effect that is the basis of the increasing numbers of emotional support animals, be they dogs, cats, or even pigs and turkeys. In one study dog owners were even found to be more likely to recover after heart attacks than were dogless people. Cat owners, on the other hand, were less likely to recover than were catless people.[3] But these types of studies are few, are correlative, and tend to consider relatively small numbers of people. Also, they do not include the other effects dogs and cats have on our lives. They do not account for the possibility that dogs and cats, just like house flies or German cockroaches, bring species into our lives, species that can make us sick and maybe even species that can make us well.

One of the species that cats bring in is the parasite *Toxoplasma gondii.*[4] *Toxoplasma gondii* is emblematic both of the ways in which species ride into our lives on our pets and of the difficulty of understanding whether our pets are good or bad for us. The story of *Toxoplasma gondii*, as it concerns us here, begins in the 1980s. A group of researchers in Glasgow was studying house mice infected with *Toxoplasma gondii.* They noticed that the infected mice seemed hyperactive when compared to uninfected mice. They wondered whether it was because of the parasite, so they gave all their mice hamster wheels to run on. The student in the group, J. Hay, counted how many times each mouse went around the wheel. During the first three days, the uninfected mice did more than 2,000 rotations on the wheel. A lot! They were hardly calm. But the infected mice did *twice* that many turns. What was more, as the days went on, the difference grew even greater. By the twenty-second day of the experiment, the infected mice were doing 13,000 turns on the wheel compared to just 4,000 turns of the uninfected mice. Here was rodential pathos of an extreme sort. Something interesting must be happening, the researchers suggested, in the brains of the infected mice. But then they went one step further. They hypothesized that perhaps the hyperactivity of the infected mice was adaptive for the survival of the parasite; perhaps the parasite caused the

**Figure 10.1** Jaroslav Flegr in his office. *(Still from* Life on Us, *directed by Annamaria Talas, credit Annamaria Talas.)*

mice to be hyperactive so that they were more likely to be eaten by a cat. The parasite *Toxoplasma gondii* can carry out the final stage of its life cycle only in cats.[5] But that was as far as the group went. They published this work and offered their hypothesis for other scholars to investigate. This was a strange enough possibility on its own. Then, ten years later, things got far stranger thanks to Jaroslav Flegr.

Flegr was born and works in Prague. In Prague, he took the necessary steps to advance his career in evolutionary biology, did some fine work, obtained a PhD, and even landed that rarest of prizes, a faculty job at Charles University. It was at Charles University that Flegr started to study parasites. Initially, he studied the parasite *Trichomonas vaginalis*, the cause of trichomoniasis. Then, beginning in 1992, Flegr became fascinated by *Toxoplasma gondii*. He began to read the work Hay had done with the hyperactive mice on wheels. When he did, he was convinced that the parasite was indeed manipulating the brains of house mice for its own ends, as Hay had hypothesized. This was happening, he thought, in houses around the world, houses in which mice ran out from beneath the stove and were pounced upon by cats, all to the advantage of the parasite. It is hard to say why Flegr was so readily convinced that Hay was right; it is even harder to say why his next thought was to wonder whether he himself, like the hyperactive mice, was also infected.

Flegr started to list for himself the ways in which his own behavior was unusual. In some real way he felt like the infected mice. It wasn't as if he was running faster than other people on a treadmill. But he did do things that, if he were a mouse, seemed more likely to get him killed and, were he living in the wild, maybe eaten by a large cat. Maybe, in addition to making mice more active, the parasite also made them less risk averse, and maybe the parasite was having the same effect in him. Once, in Kurdistan, he had found himself in a situation in which bullets were flying around him, and yet, he was not concerned about dying. At home in Prague, he was not afraid of traffic. He darted between cars, amid the orchestra of screeching brakes and honking horns, much like the infected mice dart out into the open. Nor, during the communist times, had he worried about announcing controversial ideas publicly, even with ample evidence that those who did were imprisoned or worse. How to explain it all? He must be, he began to think, infected, transformed, a man who, like Gregor in Kafka's *Metamorphosis*, was acting out a drama beyond his control.

Not long after Flegr started to have these ideas did he decide to have himself tested for exposure to *Toxoplasma gondii*. When he did, he found out that, yes, his blood contained antibodies to the parasite. He had been infected. He began to wonder which actions were his own and which were instead the impulsive gestures of the parasite. Having this very idea itself—that the parasite might be manipulating him—was reckless in a way that was emblematic of the effect the parasite might have. It was the kind of idea that was likely to marginalize him from his international colleagues. To be frank, the whole thing just seemed absurd. But he was in Prague, where wild ideas have long had a home.

By the time Flegr became interested in *Toxoplasma gondii*, scientists had begun to learn a little more about the parasite. *Toxoplasma gondii*, as Hay and colleagues noted, infects house mice (*Mus musculus*). But it also infects other household rodents, including both Norway rats (*Rattus norvegicus*) and black rats (*Rattus rattus*).[6] It can also infect geckos as well as pigs, sheep, and goats. The parasite gets into these animals when they inadvertently ingest soil or water in which its oocysts (akin to egg cases, from the ancient Greek *oon* for "egg" and *kyst* for "bag" or "bladder") are present. In the host, the beginnings of a Greek drama, or a drama described in Greek terms anyway, unfold. The stomach enzymes break down the hard wall of the oocyst, which releases the sporozoite form of the parasite (where *sporo* is from the Greek for "seed" and *zoite* the Greek for "animal") into the animal's intestines. Once there, the sporozoites invade epithelial cells inside which they turn into tachyzoites (*tachy* is from the Greek for "swift") that divide swiftly until the cells they are in die and rupture, whereupon the tachyzoites spread through the bloodstream and colonize cells in other tissues of the body. Eventually, the host immune system catches up with the parasites, and when it does, the parasites take on a new form, the bradyzoite (*brady* is from the Greek for "slow"), a form that hides within the cells of the host—in the brain, muscles, and other tissues—waiting, slowly, patiently, for the host to be eaten.

The parasite waits because in order to complete its life cycle it must be in the intestines of a cat. *Toxoplasma gondii* is a protist.[7] Like many protists, it requires very specific conditions in which to have sex and produce oocysts. It cannot have sex or produce oocysts in the soil or in rodents, geckos, or, for that matter, pigs or cows (in which it sometimes finds itself). It is picky. It finds love and fruition in the lining of felid intestinal epithelia, and only there (and they say online dating is tough). It doesn't seem to matter what kind of cat, but it must be some kind of cat. The parasite has been caught having sex in seventeen different species of felids. In this way, the life cycle of *Toxoplasma gondii* is incredibly dependent on a set of relatively unusual occurrences happening in a particular sequence. This dependence is a very important feature, a defining feature, of the parasite's life.

Once a male and a female of *Toxoplasma gondii* meet in a cat's intestine and mate, they finally produce more oocysts. The oocysts then ride the fecal highway, down the cat's gut and out into the environment. A single, small hunk of cat poop can contain twenty million oocysts. The oocysts are as persistent as seeds. They can wait for months, or even up to a year, unseen, for a mouse or other animal to ingest them. There are about a billion cats on Earth, so if only one in ten cats is parasitized and sheds *Toxoplasma gondii*, there may be as many as three hundred trillion *Toxoplasma gondii* oocysts waiting to be ingested. Conservatively speaking, there are more than 760 times as many oocysts of *Toxoplasma gondii* as there are stars in the Milky Way, a wriggling galaxy of parasites.[8]

Where and when mice, rats, and cats are abundant, such as they were around the grain silos of ancient Mesopotamia, the odds that the parasite could complete its life cycle may be high. Nonetheless, any individual lineage of the parasite able to increase its odds of success by making its intermediate host (the mouse or rat, for instance) more likely to be eaten by a cat would have an advantage. All of this had indeed been known or inferred, to some degree, by Hay, whose initial intuition about what was going on would, in subsequent years of research, prove to be right: the parasite was manipulating the mouse.

Humans, as Flegr knew when he first thought he was infected, are often exposed to *Toxoplasma gondii* in homes via cat feces. As I've noted, in nature, the oocysts of the *Toxoplasma* parasite make their way into the feces of cats and then into the soil or water, setting the stage for the cycle to begin anew. But in houses, oocysts end up in the litter box instead, sometimes in extraordinary abundances.[9] If a pregnant woman inadvertently ingests these oocysts, they break open in her stomach and the parasites divide asexually in the cells lining her intestines before spilling into her bloodstream and invading other tissues. Unfortunately, the parasite makes no distinction between the bloodstream of the mother and that of the fetus, and it travels through to the body of the fetus. Fetuses do not yet have their own immune systems. A fetus borrows antibodies from its mother, but not immune cells such as inflammatory T cells. This is a problem because *Toxoplasma gondii* is ordinarily kept somewhat in check by inflammatory T cells. As a result, *Toxoplasma gondii* can proliferate unchecked in a fetus during pregnancy, which can lead to mental retardation, deafness, seizures, and retinal damage in the fetus. (Older infections pose little risk to fetuses because parasites from older infections are likely to be established inside cells in the mother's muscle or brain, not moving through the blood.) These consequences are not common, but neither are they rare.[10] For years and years of study of *Toxoplasma gondii*, this was the end of the story: the parasite lives out a wild cycle in mice and rats and cats and, rather accidentally, poses a risk to pregnant women via cat litter.

But what Flegr also knew was that the form of the parasite to which pregnant women and other humans are exposed is the same form of the parasite to which mice and rats are exposed. At least in theory, if this parasite established itself in the cells of the brain, humans could suffer the same effects of the parasite that mice suffered. Once in the brain, the parasite could, at least theoretically, manipulate human behavior. But it seemed implausible. Whereas mice and rats are relatively small brained and so potentially can be manipulated by a tiny protist, we humans have big brains. The expansion of our frontal lobe, and the conscious thought it allows, is what makes us human and gave us the ability to invent fire, cheese curds, and computers. We have and express complex thoughts and make decisions to act. We are not simply at the mercy of our biochemistry. We are just too smart and too conscious to be controlled by the desires of a microscopic beast. Or so nearly everyone but Flegr thought.

WITH A PARASITE such as *Toxoplasma gondii*, it is hard to know just how to study what effect it might have on humans. The problem is that the way we usually study the effects of a particular pathogen or treatment is to study what it does to mice or rats, using them as model organisms. So as to avoid experimenting on humans, we poke rodents instead. The taxonomic order to which rodents belong, Rodentia, is relatively closely related to our own order, the Primates. As a result, our cells, physiology, and even immune systems are very similar to those of rodents', similar enough so that when some chemical has a particular effect on a mouse or a rat, it is very likely to have the same effect on us. Interestingly, whereas one can debate just how much dogs and cats benefit our health, there is no such debate about house mice, Norway rats, or, for that matter, fruit flies. These household animals, each of which has been inadvertently ferried around the world alongside humans, have become central to how we study our own human biology. We study them to understand ourselves; they are our mirrors. But the trouble in the context of *Toxoplasma gondii* was that we already knew that the parasite seemed to have some effect (whether it was adaptive or not) on mouse behavior. The mice were more active. It was just hard to imagine that the same would be true for us. So, what next? One could cure individuals who showed evidence of latent *Toxoplasma gondii* infections (which is to say, their immune systems showed evidence of having been exposed to the parasite), but the problem was that no one knew how to kill the slow-growing form of *Toxoplasma gondii* present inside host cells (the bradyzoite) or even how to distinguish between people with living parasites inside their cells and people whose immune systems had killed off the parasite before it was ever able to establish (but who nonetheless bore the evidence of the fight). The other problem was that Flegr really didn't have much money for any of these kinds of studies. He had his salary and he had his time. He decided to use the old approach of comparing people who were infected with people who were not. A correlation is not a causation, but it is nonetheless a starting point, a window—however muddy—into what has not been seen at all before.

The correlative study Flegr undertook was not easy, but it was cheap. He wanted to know the behaviors of large numbers of people, how they scored on personality profiles, how risk averse they were, and how likely they were to suffer problems associated with risky behavior (such as getting into car accidents). He went door to door, like a medieval salesman, hawking wild ideas and blood tests. He didn't go out around Prague but instead kept things simpler. He just went through the halls of his own university department. As he would report in his paper, most of his participants were the faculty, staff, and students of the Faculty of Science at Charles University. He asked his coworkers, 195 men and 143 women in total, the 187 questions of the Cattell's 16 Personality Factors (PF) survey, an assessment developed in the 1940s and used around the world to assess the magnitude of sixteen different personality factors, including warmth, liveliness, social boldness, and dominance. With the exception of Flegr and his collaborator (both of whom also participated in the study), none of the other participants knew whether they had been infected with *Toxoplasma gondii* before answering the questions. In addition to the personality test Flegr also gave each participant a skin test for *Toxoplasma gondii*. Each participant was given an injection of *Toxoplasma gondii* antigen, and if an immune reaction caused a small bump at the injection site after forty-eight hours, the participant was considered to have been at some point in their life infected with *Toxoplasma gondii*.[11] This didn't necessarily mean that the participant still had *Toxoplasma gondii* in their body, or even that the parasite had ever invaded their cells, just that at some point they had ingested the parasite in sufficient abundance so as to cause their immune system to try to fight it off. The work took fourteen months during 1992 and 1993. Flegr's colleagues at Charles University thought he was quirky, but nonetheless, they consented to participate in the study (and, in doing so, disclosed many details about their lives).

When Flegr considered the data, he saw that men who had been exposed to *Toxoplasma gondii*—men like him—were different from those who were not. The men who were infected were more likely to be risk takers (to have a high "social boldness" factor on the test), and so to disregard rules and make rapid, and potentially dangerous, decisions. Overall, for both sexes, the personality types of those who were infected differed from those who weren't. When Flegr looked more closely at the data, they seemed to explain key features of the human world to him. Inasmuch as the participants were his colleagues, the results explained his world. For example, twenty-nine of his fellow professors tested negative for *Toxoplasma gondii*. Those people tended to be leaders—people who made slow, considered decisions. Ten of the twenty-nine had been department heads, vice deans, or deans. Conversely, only one of the infected professors had ever had a leadership role (as a department head).[12] Subsequent studies would show similar patterns. For instance, Flegr found that individuals infected by *Toxoplasma gondii* were two and a half times more likely to get into car accidents (a finding later repeated in two independent studies by Turkish research groups, a study in Mexico and a study in Russia).[13]

He was emboldened.[14] He advocated more strongly for his ideas. There was something there, but he knew what people would say. They would argue that people who were exposed to *Toxoplasma gondii* were just different to start with, that risky people were just more likely to get the parasite, for example. He couldn't rule this out, not in a formal way, but he also couldn't imagine why being risky would make you more likely to be exposed to a parasite found in cat feces. The idea that risky people were more likely to have cats or to inadvertently ingest cat feces seemed like a stretch.[15] Then again, so was his idea.

We don't know when humans were first exposed to *Toxoplasma gondii*. In one possible scenario, our exposures to the parasite were relatively rare until the origin of agriculture. With the advent of farming, we began to store grains. Our grain stores fed large populations of grain-feeding insects and mice (*Mus musculus*). Grain was money, and mice were eating the money.[16] As mice populations grew, so too did cat populations, and cats were domesticated so that agriculturalists could more persistently avail themselves of the cats' services. Once cats were domesticated, our exposure to their feces increased. So too did the frequency with which we were exposed to *Toxoplasma gondii*.[17] By 7500 BCE, a cat was buried in a shallow pit beside a human in Cyprus. The cat was not chopped into pieces. It was not cooked. It appears to have been curled neatly, as many cultures curl their human dead. Cats are not native to the island of Cyprus, so this cat (or its ancestors) must have come to the island with humans in a boat. The human beside the cat was buried with jewels and ornaments; he was powerful and wealthy. This burial implies our relationship with cats has long included an element of reverence, or at least appreciation.[18] This cat, in particular, was probably already domesticated (though it is hard to know for sure on the basis of the bones alone).

Our first encounter with *Toxoplasma gondii* may have been in an early agricultural settlement such as one of those on Cyprus. The man buried beside his cat and the cat might both have been infected. The other possibility is that we began to be exposed to *Toxoplasma gondii* even earlier in human prehistory. As hunter-gatherers, we might have ingested it, accidentally, from the soil, much as would a mouse. Or we might have ingested it in uncooked meat (this is the other way we can be exposed to the parasite, by eating meat, such as that of pigs or sheep, in the cells of which the parasite dwells). Then, because our ancestors were not infrequently eaten by large cats, every so often we aided and abetted the parasite in its quest to get to its favored final destination. Our ancestors, particularly as young children, were eaten by cats more often than we might imagine. But even if this latter scenario is correct, it still seems likely that with the dawn of agriculture and the welcoming of cats into our homes, the frequency of our interactions with—and infections by—the parasite must have increased. Either way, if Flegr's hypothesis was right, the parasite might have been affecting our behavior for a very long time. He was, in other words, considering not just how this parasite affects us now but also how it may have been affecting our ancestors for generations. One wonders, for example, whether Genghis Khan was infected by *Toxoplasma gondii*. Or maybe Columbus.

During the years in which Flegr considered the many ways *Toxoplasma gondii* might affect humans, and has affected humanity, a few other biologists quietly continued to unravel the effects the parasite had on rodents. One of them was Joanne Webster, a specialist on pathogens spread by nonhuman animals. (Webster calls herself a zoonotic epidemiologist.) Like Flegr, Webster had also decided to follow up on Hay's experiments in Edinburgh. But unlike Flegr, Webster would do experiments. Hay had worked with house mice. Webster studied Norway rats, lab rats. In Norway rats, just as in house mice, *Toxoplasma gondii* divides asexually in the bloodstream and spreads through the body, inserting itself into the cells of muscles such as the heart as well as into the cells of the brain. When inside the cells of the brain, the parasite encysts and can stay encysted for years, for the lifetime of the host even. Webster was able to show in one careful experiment after another that when rats are infected with this parasite, the rats become more active, just as do house mice.[19] Rats also become unafraid of the ordinarily terrifying smell of cat pee, a change that, just like the hyperactivity of mice, made the rats more likely to get eaten by cats.[20] Nature can make ants love beetles; it can also make mice and rats walk right into the gaping mouths of their predators.

Slowly, Webster began to understand how the parasite causes all of these changes in the rats. It appears that once it arrives in the brain, the parasite produces the precursor to dopamine,[21] which, together with other chemicals and mechanisms we have yet to understand, causes mice and rats to be more active, to be less afraid of cat pee, and to be more likely to be eaten by a cat. Because the species cats eat can live both indoors and outdoors, both indoor and outdoor cats are hosts to *Toxoplasma gondii*.[22]

Webster's work helped to launch the study of the ways in which many parasites, not just *Toxoplasma gondii*, manipulate their host's behavior. We now know such parasitic manipulation to be common. Fungi manipulate ant brains, wasps manipulate spiders, tapeworms manipulate isopods, and so on. But, apart from Flegr, no one who studied *Toxoplasma gondii*, including Webster, focused on the ways that it might influence humans.

Webster's job offered her the scope to potentially study *Toxoplasma* and humans. One of Webster's faculty appointments is in the School of Medicine at Imperial College, so she works every week among colleagues focused on human diseases. But the sort of correlative work Flegr was doing was unconvincing to her colleagues, as would be any similar work Webster might do. Webster didn't necessarily have to convince her colleagues that the work was interesting or that the results were significant to continue, but it would help. Academia is built on a culture of respect gained and, just as readily, lost. If you lose the respect of your peers for your work, you lose their support, you lose their collaboration, you lose, too, potentially, any favors you might require of them in the future (and academics seem, nearly always, to be in need of favors). But just as problematic as the fact that the work wouldn't convince Webster's colleagues was that it wouldn't fully convince her either. Her training was in experiments, in testing hypotheses in the lab, and there weren't many aspects of the story of *Toxoplasma* and humans that could be studied experimentally. She couldn't ethically give people *Toxoplasma gondii*, and no one knew how to get rid of it once it was established inside cells (so she couldn't cure people and look at the effects). However, as Webster continued her work, she spotted a possibility. Along with many other phenomena, Flegr hypothesized that the parasite might influence not only behavior but also psychological health. More specifically, building on the work of Flegr, E. Fuller Torrey, a psychiatrist at the Stanley Medical Research Institute, and Robert Yolken, a professor of pediatrics at Johns Hopkins University Medical Center, suggested that *Toxoplasma gondii* might be partially or even wholly to blame for schizophrenia.[23] Both schizophrenia and *Toxoplasma gondii* tended to be clumped within particular families, but not in a way that seemed to be exclusively genetic (rather that it was more related to the houses people lived in than the genes they had). In addition, the drug used to manage the symptoms of schizophrenia appears to sometimes rid patients of the *Toxoplasma gondii* hiding in their cells. Presented with these observations, Webster had an idea. She wondered whether the way in which the schizophrenia medications worked was by suppressing or even killing *Toxoplasma gondii*.

Webster did an experiment. It is what she does. She orally infected forty-nine rats with *Toxoplasma gondii*. She then pretended to infect an additional thirty-nine control rats by giving them an oral dose of what was effectively saltwater. Each group of rats, infected and control, was then further divided into four groups. One group received no additional treatment, one received valproic acid (a mood stabilizer), one received haloperidol (an antipsychotic), and the last group received pyrimethamine, a drug known to kill parasites, including, under some conditions, *Toxoplasma gondii*. Afterward, she put the rats, one by one, into a one meter by one meter square pen. In each corner of the pen she put fifteen drops of one of four odors. In one corner, she put wood chips soaked with fifteen drops of the rats' own smell, rat urine. In another corner, she put wood chips soaked in a neutral smell, water. In the third corner, she put wood chips with rabbit urine, with the idea that rabbit urine should have no specific effect on the rats because rats have no reason to be scared of or attracted to rabbits. And in the final corner, Webster set down wood chips soaked with cat urine. Webster works at one of the most prestigious universities in the world. She has made major discoveries, and yet here she was, day after day, laying out the urine portfolio. After the pen was set up, an individual rat was released and Webster or someone from her team watched and noted the proportion of time the rat spent in each corner. This was done again and again and again for a total of 444 hours of observation of eighty-eight rats. When the data from these observations were tallied, they totaled 260,462 lines. Webster took the time to count. Those lines of data showed that the uninfected rats spent more of their time huddled near the familiar and "safe" smell of their own pee or that of other innocuous animals such as rabbits. The uninfected rats wisely avoided the cat pee areas. The infected, but unmedicated rats behaved differently. They entered the cat pee corner more often, and once there they tended to stay, as if completely unaware of the potential danger the pee might signal. Amazingly, the rats infected with *Toxoplasma gondii* but treated with either of the schizophrenia medicines or the antiparasite medicine acted more like uninfected mice. Compared to the rats infected with *Toxoplasma gondii*, but not treated with any medicine, they were less likely to enter the cat pee area and, if they did, they didn't stay as long. They were, to use a loaded term, cured.[24]

Webster published her paper on schizophrenia, schizophrenia medicines, and *Toxoplasma gondii* in 2006. The work was compelling but still confined to mice, not humans. A human study needed to be done, but it couldn't be just correlational (at least not for Webster). Nor could it be experimental. There was, though, a third option, a longitudinal study. Someone could track people through time to see whether individuals infected with *Toxoplasma gondii* were more likely, over years, to develop schizophrenia as compared to uninfected (but otherwise similar) individuals. This wasn't Webster's kind of study; it was not the sort of science she did, and yet, if someone were to find a way to do such work, it would be an elegant test, a test that would extend her work in a way that might finally get the attention of doctors. It was hard to imagine who might have the right sort of data. Such a data set would need to include not only health data from different time points but also blood samples from those time points. Among the only groups in the world with such data was the United States military.

The US military collects health data on all recruits. It also collects samples of their blood. Another epidemiologist, David Niebhur, at the Walter Reed Army Institute of Research, decided to study these data to see whether schizophrenia was, indeed, associated with *Toxoplasma gondii* infection. Niebhur went through the military database and found 180 service members who had been medically discharged from the army, navy, or air force between 1992 and 2001 because of a schizophrenia diagnosis. In the database, Niebhur and his colleagues then found another 3 soldiers without schizophrenia for each person diagnosed with schizophrenia. These control individuals matched the patients diagnosed with schizophrenia in terms of age, sex, race, and branch of military service. The researchers examined the blood serum samples collected by the military to see whether individuals who developed schizophrenia were more likely than controls to have been infected with *Toxoplasma gondii* prior to the onset of schizophrenia. They were. The soldiers who were discharged with schizophrenia were significantly more likely to have tested positive for exposure to *Toxoplasma gondii* than those solders who were not diagnosed with schizophrenia.[25] Niebhur and his colleagues found that people exposed to *Toxoplasma gondii* have a 24 percent higher risk of developing schizophrenia at some point in their lives than people who have never been exposed. If you have been infected by *Toxoplasma gondii*, ever, your risk of schizophrenia is at least 24 percent greater than that of someone who has not been. Time has added nuance to the work led by Niebhur and his colleagues, as has replication. The number of papers published on the parasite has increased. To date, fifty-four studies have looked at links between schizophrenia and *Toxoplasma gondii*. All but five have found evidence of such links that suggest that infection by *Toxoplasma gondii* increases the risk of schizophrenia.[26]

Stepping back, it now seems Flegr was on the right path. *Toxoplasma gondii* appears to act in our brains much as it does in the brains of mice and rats. Nor are we the only primates affected. A recent study has shown that infection by *Toxoplasma gondii* appears to cause chimpanzees, our close relatives, to be attracted to the smell of cat pee, specifically leopard urine.[27] Infected humans, or at least infected human men, are also more likely to regard the smell of cat urine as pleasant than are uninfected men.[28]

The proportion of people who have been infected by *Toxoplasma gondii* is huge. Some of these infections resulted from eating meat that was not fully cooked, meat in which the parasite lurks, wriggly, in muscle cells. But many are from our cats. Just how common are *Toxoplasma gondii* infections? In France, upward of 50 percent of all people show evidence of latent infections. This parasite could explain the behavior of much of a nation. It isn't culture that makes the French enjoy red wine, meat, and cigarettes so much—it's just that their parasites give no shit about risks. But lest the non-French get too haughty, I should note that the infection rate in other countries is also high. Forty percent of Germans have been infected. In the United States, upward of 20 percent of all adults have been infected by *Toxoplasma gondii*. Globally, more than two billion people have been infected at some point in their life.[29]

In and of itself, the story of *Toxoplasma gondii* is a big deal. *Toxoplasma gondii* may well be the most common parasite in humans, or at least the most common parasite with a big effect. Face mites, which we study in my lab, are more common than is *Toxoplasma gondii* (all adults we have ever sampled have face mites),[30] but face mites don't seem to cause any negative effects. Of parasites with negative effects *Toxoplasma gondii*, long neglected, appears to be king of commonness. But the story of mice, cats, *Toxoplasma gondii*, and the other parasites of cats is really a broader lesson about the complexities of opening our doors to domestic animals. Collectively, we seem to have eagerly decided that insects and microbes in our homes are bad, but our pets are good. Yet, when we let a cat in through our front door, the *Toxoplasma* hiding inside that cat comes in too. *Toxoplasma gondii* does not travel alone. Dozens of other species, all of them more poorly studied than it, also come inside with our feline friends. And lest cat people, be they ordinary women or men strangely drawn to the smell of cat pee, feel singled out, very similar things are happening with the other kinds of domestic animals we allow indoors.

Over the last twelve thousand years, we have welcomed in many domestic animals, be they cats, ferrets, dogs, guinea pigs, or comfort ducks. Each brought with it other species. Cats brought *Toxoplasma gondii*. Guinea pigs appear to have brought human fleas. But dogs, oh dogs, dogs are a veritable smorgasbord of worms, insects, bacteria, and more.

Seven years ago, I had the idea to have students in my lab compile a database of all the parasites associated with each kind of domestic animal. The idea was to make a complete list, pet by pet. Meredith

Spence was the student charged with cataloging the species that live on domestic dogs. I imagined that after Meredith finished with dogs, someone else would do cats, another student would work on rabbits, and so on. We'd work our way to camels and guinea pigs. We never got beyond dogs. Meredith spent a year compiling the list for dogs, then two years, then three. Eventually, she graduated from North Carolina State University with an undergraduate degree, went off to work in a veterinary clinic, came back to the university to start graduate school, and is now almost finished with her PhD. She continues to compile the list of species that live on and in dogs.[31] It includes some species you might expect—dog fleas, of course, and the *Bartonella* parasites that ride in fleas.[32] In addition to dog fleas, it includes up to six other species of fleas. It includes mites of the genus *Demodex*, which are relatives of the mites that live on the faces of all adult humans, as well as several other species of mites. It includes a dozen different species of ticks, some big, some small. It includes dog-chewing lice and two other kinds of lice. Meredith's list includes all of these species. It also includes an entire circus of gorgon-headed worms. In total there are more than twenty kinds of worms, some of which have never really been studied in any detail, including tapeworms of the genus *Echinococcus*.

The story of *Echinococcus* tapeworms is just starting to be resolved. We are currently, in our understanding of *Echinococcus*, only as far along as we were with *Toxoplasma gondii* in 1980. The study of *Echinococcus* is just beginning. Most tapeworm species have as their final hosts carnivores—whether those are dogs, cats, or sharks—but can be picky, just as is *Toxoplasma gondii*, with regard to just which carnivore species. Adult *Echinococcus* tapeworms greatly prefer dogs. Dogs are the "final" or "definitive" host of *Echinococcus* tapeworms, which is a boring scientific way of saying dog guts are "the place that the worms go to have sex, produce eggs, and die." One might imagine that because dogs are, like cats, carnivores, *Echinococcus* tapeworms might be able to mate in cats. But they can't (much as *Toxoplasma gondii* is unable to mate in dogs). To this particular parasite, something about a dog's gut is just right.

Once two *Echinococcus* tapeworms have mated inside a dog, the resulting eggs leave the body of the dog in its feces. Having been deposited, they wait. Grazing animals often ingest a little dog feces, inadvertently, along with their grass. This is one of nature's less known realities. The grazers are often goats or sheep. Though where goats and sheep are rare, a deer or even a wallaby will do. In the stomachs of the grazers, *Echinococcus* eggs hatch. The newborn larvae spread through the body of the animal and nestle there in cysts in organs or even in bones. When a grazing animal dies, the dogs are then exposed to these parasites if and when they eat any of the cysts. In the same way, humans can also be infected by the larval *Echinococcus* tapeworms when eating grazing animals such as sheep. The larval tapeworms can form cysts inside humans much as they might inside sheep, except that inside humans the cysts never stop growing; they can reach the size of basketballs. Eating an infected sheep is the more "glamorous" way to develop an *Echinococcus* cyst in your body. The less glamorous way is by accidentally ingesting a little dog feces that contain the eggs, which happens far more often than you might hope, such as when people let their dogs lick their faces. The world is vulgar and alive.

The story of the *Echnicoccus* parasite begs questions. Does this parasite manipulate the sheep that are infected or the humans who are infected so as to make them more attracted to dogs? Do dog people love dogs because they are possessed by the biochemistry of the worm? No one knows; stranger things, as should be clear by now, happen in the wilds of our daily lives.

Some parasites and pathogens of dogs, such as the rabies virus, are common in some regions (or times) but rare in most places, at least today. *Echinococcus* was one of the most common species in dogs on the list of dog parasites that Meredith Spence compiled and it was common in many regions, but it wasn't as common as heartworms (*Dirofilaria immitis*). Meredith Spence now studies dog heartworms; it is the project to which her initial cataloging helped lead her. Heartworms are nematodes. They invade and live in the living hearts and pulmonary arteries of dogs, where they ultimately grow so dense as to clog the normal movement of blood. In the United States, up to 1 percent of dogs have been infected with heartworms. In some countries, more than half of dogs have been infected. The heartworm gets into dogs via mosquitoes. The worms ride inside mosquitoes. Then, in those moments when mosquitoes bite dogs, the worms quickly swim down and out of the proboscis of the mosquito into the wound left by its bite. From the wound, they crawl into subcutaneous tissue of the dog. From the subcutaneous tissue, the worms migrate through the muscle fibers before spilling into blood vessels en route to the heart. By the time they reach the heart, the worms have molted several times and are adults. Ever the romantics, these adult worms mate inside the heart. The evolution of dog heartworms has never really been studied in much detail, nor has that of the many other species of heartworm. Meredith is focused on the mosquito side of the story, so the evolutionary story of the heartworms won't likely attract her attention anytime soon. Here, then, is a beautiful project for those who might be interested (my guess is that new species of heartworms are common, unnamed, and riding around your neighborhood right now in mosquitoes). Dog heartworms do not usually invade the hearts of humans. It happens rarely enough (hundreds, not thousands, of cases a year) that when it does, the doctors who have spotted the problem gather around and snap selfies with the afflicted patient. In only one case have heartworms been found mating in a human heart; most of them get stuck, during their corporeal peregrinations, in the pulmonary arteries, where, unable to move forward or back, they die. More rarely, a worm gets stuck and dies in the blood vessels of the eyes, brain, or testicles. Again, though, these cases are rare.[33]

What is not rare is the exposure of humans to heartworms. Many people show antibodies for dog heartworms, which is to say many (perhaps most) humans have been bitten at some point by a mosquito carrying dog heartworms. The worms tunneled into their (perhaps your) skin, but the human immune system killed the worms. When this happens, the person the worm has tried, unsuccessfully, to invade notices nothing different about their life. However, recent research suggests that even a single exposure to these worms can alter the immune health of people, favoring antibodies that predispose people to asthma relative to those who haven't been exposed. In other words, it may be the case that a mosquito, in landing on you, gives you a worm that your immune system kills but that leaves a legacy, a kind of ghost, in the form of an increased predisposition to sneezing, coughing, and wheezing.[34] The fact that we are often being bitten by mosquitoes carrying dog heartworms is largely a function of having let dogs into our lives. The worms are present in our environment because the dogs are present (they can also be present where coyotes or wolves are present, but few are the neighborhoods where wolves and coyotes outnumber dogs). You don't even have to have a dog yourself to have the worms slide into your body. It is enough for there to be dogs present in your neighborhood. As many as twenty other parasites are common in dogs, connections to their wild wolf ancestry and to the world outside your home. What's more, as Meredith's cataloging has revealed, dozens of other parasites are at least occasionally found in domestic dogs.

I find the basic biology of *Toxoplasma gondii*, *Echinococcus* tapeworms, and heartworms to be wondrous and fascinating. But, like anyone else, I'd prefer to avoid being infected. Opening the door to a cat or a dog increases the risk of infection. Fortunately, the strongest negative effects of these infections—be they schizophrenia, a tapeworm, or dead heartworms in your testicles—are rare in most regions. Also, some of the risks posed by dogs and cats can be ameliorated by preventive measures dog and cat owners can take. For example, heartworm medicines reduce the abundance of heartworms in dog populations (though their use also increases the speed at which heartworms resistant to those same medicines evolve). Others, though, such as those posed by *Toxoplasma gondii*, can't, at least not yet anyway.

I don't propose to have the answer for how to balance the benefits of having a pet relative to the most common kinds of costs. The answer ultimately depends on where and how we live. In some regions, cats still help protect grain from mice and rats. In some regions, dogs still help shepherds work and protect sheep. But in a modern Western context, what these animals most often offer is companionship. Their value as companions increases in proportion to our need for company, in proportion even to our loneliness and despair. The more urban and isolated we become, the more likely they are to provide such benefits. Also, the more urban and isolated from nature we become, the more likely dogs and cats may provide another, new kind of benefit, that of connecting us to beneficial species.

We first considered the beneficial effect of pets on the bacterial life in houses when we surveyed forty homes across Raleigh and Durham, North Carolina. One of the questions we asked participants in that study was whether they had a dog. Almost 40 percent of the variation from one home to the next in terms of species of bacteria in the home resulted from the presence or absence of a dog.[35] This was a huge effect. The effect of the dog was in part the result of a set of soil microbes that were more common in houses with dogs. We imagined that the dogs were just carrying these soil microbes in from the outside, but a recent study found soil microbes living in the fur of many different mammal species.[36] It is possible that the normal fur microbiome for many mammals and the normal soil microbiome overlap. In addition to soil bacteria, dogs also left drool-associated bacteria around homes as well as a few fecal bacteria common in dogs, but not as common in humans (and so identifiable in the mix).

Once we had data from a thousand homes, we were able to consider whether cats had an effect on the bacteria in homes. They also do. For reasons we don't totally understand, some species of bacteria, including some insect-associated bacteria, become rarer when a cat is in a home.[37] Maybe the pesticides put on cats in the form of flea collars, drops, and powders kill the insects, which in turn kills their bacteria (though we'd expect this to be the case for dogs too). Maybe cats eat the insects (and kill their bacteria). Still, cats provided a way for hundreds of bacterial species to get into our houses. Most of these species, just as with dogs, appear to be associated with the bodies of cats—their skin, their fur, their feces, their saliva. What cats don't seem to bring in with them are soil microbes. Perhaps it is because cats are smaller, perhaps it is because they clean their feet. We don't know.

I suspect that at many points in our history the average bacterial species that a dog or cat brought in would have been, like the protists or worms, reasonably likely to have a negative effect on us if it had an effect at all. But our moment is an unusual one. Today, as outlined in the biodiversity hypothesis, in much of the world we are as likely to be sick from the bacteria we don't have as from the bacteria or parasites we do. It is possible that, for children who fail to be exposed to enough of the right bacteria, exposure to what the dog or cat brings in offers some of the same benefits as sniffing the biodiverse dust from Amish houses. Recent studies suggest that having a dog does tend to lead to a reduced human risk of developing allergies, eczema, and dermatitis, particularly for children who are born into a household with those animals. The most comprehensive review of the literature found that children who live with pets tend to be less likely to suffer from atopic dermatitis.[38] A similar study in Europe found the same result for allergies—having a pet reduced the owner's risk of allergies, more in some regions than in others.[39] Across studies, cats tend to show effects similar to those of dogs, but the effects tend to be weaker and less consistent.[40]

In parts of the world where we have become distant from wild biodiversity, dogs and cats may be good for our immune systems. The effect of dogs and cats on our immune systems could work in two ways. It may be that the bacterial species dogs and cats bring in compensate for the exposures we no longer receive. We live lives so disconnected from biological diversity that even being exposed to a little gunk from a dog's foot can be an improvement. Alternatively, our children may inadvertently be benefiting from the fecal bacteria of dogs and cats in their own guts. Children with dogs in their homes tend to acquire dog gut bacteria by eating food from the ground with dog feces on it or by being "kissed" by a dog that has just kissed the backside of another dog.[41] It is possible that what dogs (and maybe to a lesser extent cats) offer is not some generalized exposure to bacterial biodiversity but instead a chance to pick up our necessary gut bacteria when they have gone missing. It is now well documented that when specific gut bacteria are absent, their absence can cause a variety of health problems (from Crohn's disease to inflammatory bowel disease and more). If this feces-eating hypothesis is right, we would expect C-section babies, who often fail to get all of the bacteria that they need,[42] to get more benefit from dogs. They seem to. We would also expect that in houses where other sources of fecal microbes are more available, such as through interactions with dirty-fingered siblings, the effect of dogs would be less pronounced. This seems to be the case too. Dogs have less of an effect on the allergies and asthma of children with siblings. In general, I think the evidence so far supports the idea that dogs can benefit us by bringing in soil bacterial biodiversity and by offering up fecal microbes we have lost, but these services are only benefits because we now live in a world disconnected from wild nature, so diconnected that adding a little dog dirt and feces to our lives is a kind of solution. If we put these results back together with the stories of *Echinococcus*, and the heartworm, the results of having a dog seem likely to differ depending on just which species dogs bring in, whether the species are bacteria or worms, and if worms, which worms. Sometimes, when we are trying to figure out how to make simple decisions to better our lives, biodiversity is a real jerk in its complexity.

The truth is we don't really know yet what the average consequences are of letting a dog or a cat, much less a ferret, tiny pig, or turtle, into our homes. And if we struggle to figure out whether dogs and cats make us healthy, you can get a sense of why it is also hard to figure out just which of the hundred thousand or so bacterial species from other sources sometimes found in our homes or on our bodies are the ones we want. But that hasn't stopped people from trying. In fact, at one point in the 1960s it seemed as though doctors might soon plant gardens of bacteria on the bodies of babies across the United States, and perhaps in hospitals and houses too. Then they did.

# 11

# GARDENING THE BODIES OF BABIES

We will now discuss in a little more detail the Struggle for Existence.

—CHARLES DARWIN

Sweet flowers are slow and weeds make haste.

—WILLIAM SHAKESPEARE

We dream about progress. When we do, we imagine progress to be technological. We imagine the past to be lesser than the present, which is lesser still than the future. But when it comes to our management of the life around us, particularly the life in our homes, this may not be the case. Where we have controlled dangerous pathogens, it has been a huge step forward, but we went too far and killed the beneficial species too. Then we inadvertently built our houses in ways that favor problem species—fungi living in our walls, new pathogens lurking in our showerheads, and German cockroaches running beneath our doors. Meanwhile, the entire time there was always another way, another road. We could have, years ago, figured out how to favor species that benefit us in our homes. This may seem like a dangerous proposition. But it is less dangerous than the world we have made. What is more, it is a method that has already been tried. It was tried, in all places, on the skin of newborn babies. And it worked.

It all began in the late 1950s. A pathogen called *Staphylococcus aureus* type 80/81was spreading rapidly among hospitals in the United States.[1] It threatened people who visited the hospitals and then, when they went home, threatened their families too. It was especially dangerous to babies, for whom it was, as one study at the time noted, "responsible for more potentially serious infections in hospitals than any other microbe."[2]

*Staphylococcus aureus* type 80/81 (hereafter just "80/81") lodged itself in a person's nose or belly button and, having done so, became essentially impossible to eradicate. It was resistant to the main antibiotic used at the time, penicillin. Penicillin was first available to the public in 1944. Penicillin didn't work for all kinds of pathogens (controlling the tuberculosis bacterium, *Mycobacterium tuberculosis*, for example, awaited the discovery of a new kind of antibiotic, streptomycin, produced by *Streptomyces* bacteria). But it did work on pathogenic strains of *Staphylococcus aureus* until, that is, strain 80/81 evolved. This strain was no longer killed by penicillin.[3] Worse yet, it spread alarmingly quickly.

By 1959, Presbyterian Weill Cornell Hospital in New York was among the many hospitals in which 80/81 had become common in nurseries. In this, the hospital was ordinary. What was different about Presbyterian Weill Cornell was that two people—Heinz Eichenwald and Henry Shinefield—decided to find a solution to the problem of 80/81.[4] Eichenwald was a doctor in the Department of Pediatric Diseases at the Cornell Medical Center at Cornell University, and Shinefield was a newly appointed assistant professor in the same department. In their work together, the two men would usher in an entirely new kind of medicine and an entirely new way of managing the life indoors.

Eichenwald and Shinefield diligently studied the nurseries in Presbyterian Weill Cornell. Every day before they went home, they checked the nurseries in the hospital for the presence of 80/81. They couldn't have quite described what they were searching for; they would know it when they saw it. It was tedious work, but the tedium became a kind of ritual. Then the ritual began to pay off with novel observations.

The first observation was that the nurseries at Presbyterian Weill Cornell with the most infections had all been visited by the same nurse, a nurse later confirmed to have 80/81 lodged in her nose (hereafter we'll just call her "Nurse 80/81"). Infections seemed to follow Nurse 80/81 wherever she went, or nearly so. Nurse 80/81 was clearly to blame. Infections in hospital nurseries were very common; so too were infectious nurses. In most hospitals this nurse would have been dismissed and the story of her effects on the nursery would be over. Case closed. Initially, this is what happened. Nurse 80/81 was, as Shinefield and Eichenwald would later note, "removed." But there was more to the story at Presbyterian Weill Cornell.

Nurse 80/81 had contact with a total of 68 babies, 37 on the day of their birth and 31 not until twenty-four hours after their birth, on their second day of life. Of the 37 babies that she handled in the first twenty-four hours of their lives, one-fourth were colonized by 80/81. However, of the 31 newborns that she handled after they had been alive for twenty-four hours, none were colonized by 80/81. Their noses were, instead, colonized by other bacterial strains, including other apparently harmless strains of *Staphylococcus aureus*. Herein was a mystery among the bodies and fates of newborns. Why did the babies who were held by Nurse 80/81 during the first twenty-four hours of their lives get colonized by 80/81, even though those who were held when they were just one day older did not? In comparing these two groups of babies, Eichenwald and Shinefield developed a hunch about what might be happening. A hunch can make a career; it can also ruin one.[5]

Eichenwald and Shinefield imagined two possible explanations for the pattern they observed. The first and more ordinary explanation was that age conferred some sort of immunological maturity that better enabled the newborns to defend themselves. The older babies simply rejected the 80/81. Their bodies killed the pathogen before it could establish itself. Let's call this the tough baby hypothesis. Scientists aren't supposed to discount hypotheses that they find boring and unfortunate, but they do; and this is how Eichenwald and Shinefield felt about the tough baby hypothesis. It was boring.

Eichenwald and Shinefield's second hypothesis was outlandish and slightly wild but also way more interesting. Eichenwald and Shinefield wondered whether the older babies had had more of a chance to be colonized by other strains. "Good" strains of *Staphylococcus aureus* might confer resistance to newly arriving pathogens (such as 80/81) much as if they were a kind of force field. If the latter hypothesis, which Shinefield termed "bacterial interference," was right, it suggested a whole new world in which beneficial bacteria might be seeded onto bodies as well as hospital surfaces and homes.

In developing these ideas, the men knew, as they would note, "colonization of the newborn with *Staphylococcus aureus* is a normal event," an event that will "always take place sooner or later."[6] This was well established. A number of studies had, by then, shown the skin of healthy adults hosted a shag carpet–like layer of microbes. In the nose, belly button, and a few other spots, those microbes nearly always included species of *Staphylococcus aureus* living in dense biofilms. Other patches of skin—be they forearms or backs—were dominated by other species of *Staphylococcus*, species of *Corynebacterium*, *Micrococcus*, as well as other oligarchs of the flesh.[7] It is the normal condition of mammals to be covered in a shaggy layer of bacteria (though we now know that just which species form this layer depends greatly on the identity of the mammal). Even when naked, we are cloaked, and the same is true for the surfaces in our houses. We also know that babies, in utero, do not have microbes on their skin (or in their guts or lungs); babies' bodies are colonized during birth.

In this context, Eichenwald and Shinefield imagined that the cloak of newly established microbes on the skin of these day-old newborns, particularly in their noses and belly buttons, might prevent other microbes from colonizing or thriving. More specifically, they thought that beneficial strains of *Staphylococcus aureus* might outcompete pathogens by taking up space and food resources before the pathogens could gain a foothold.[8] Ecologists call this scenario "exploitative competition." It was also possible that in addition to preventing the establishment of pathogens through exploitative competition, the established species might produce kinds of antibiotics, called "bacteriocins," that actively deter or even kill other, later-arriving bacteria.[9] Ecologists call this "interference competition."[10] Both kinds of competition are common in nature and well documented out among the plants in grasslands or ants in rain forests, but the idea that they might be occurring among bacteria on bodies and in buildings was radical at the time. It was not without precedent, and yet even so, it was a marginal concept—not so much lunacy as heresy.

At that time, medicine, particularly medicine dealing with infection, was focused on killing bad species or strains when they began to cause problems. It had been so ever since Snow found the contaminated well in Soho, London, ever since Louis Pasteur worked out that individual pathogenic species can cause disease (the germ theory). Almost no one searched for beneficial species or considered that illness might sometimes result from their absence.[11] The focus was on pathogens. Pathogens and how to kill them. The culture was akin to that before the domestication of wild animals, a time when our ability to deal with large beasts relied entirely on our ability to avoid them or kill them. Eichenwald and Shinefield thought differently. They imagined that effective medicine, and human health more generally, could rely on a more holistic view of life.

They, along with their colleague John Ribble, devised an experiment. They wanted to see what would happen when newborns initially placed in a nursery with virtually no 80/81 were moved to the nursery in which more than half of all newborns were infected. Would the transferred babies be protected by their early colonization by bacteria other than 80/81? This experiment was carried out at Presbyterian Weill Cornell. The babies were put in a safe nursery, a nursery without 80/81, for sixteen hours. They were then transferred to a nursery in which 80/81 was not only present but ubiquitous. The result was unambiguous. The babies that started in the nursery without 80/81 were protected from 80/81, even though they were just a day old.[12]

This experiment was clever (though ethically dubious). It suggested a role for beneficial bacteria as a defense against pathogens; the beneficial bacteria appeared to outcompete or even to kill the pathogens. Yet it left open a range of other possible explanations, not least of which was the oh-so-very-boring tough baby hypothesis. Eichenwald and Shinefield decided to do the perfect experiment; they decided to garden the bodies of babies, to focus on intentionally favoring beneficial species rather than simply disfavoring pathogens.

The gardening would be carried out using a bacterial strain that Shinefield had isolated from another nurse, a nurse named Caroline Dittmar, who had visited the nursery in which the babies were not colonized by 80/81. Dittmar's nose was colonized by a strain called *Staphylococcus aureus* 502A. Strain 502A was the same one found in forty of the babies in the healthy nursery, a strain Shinefield and Eichenwald believed to be both safe and capable of interference. The men studied Dittmar's 502A for two years. It didn't seem to be associated with disease of any kind, either in babies or in their families. It would later be revealed that the reason Dittmar's 502A does not cause infections is because it lacks the ability to penetrate the mucosa of the nose to get into the bloodstream. When it finds a way into the bloodstream, it is just as pathogenic as any other bacterial species.[13] Even while they were still studying 502A, Eichenwald and Shinefield began to use it to inoculate babies. They started with lower densities of the strain, and then, once it was clear that more was needed to get the bacterium to "take," they used higher densities of about five hundred individual bacterial cells.[14] Up to one year later, 502A still seemed to be present in the noses of most of the babies that had been inoculated (fewer in the belly buttons, for reasons that can only be guessed). What was more, the mothers of the babies also began to be colonized by 502A.[15] Whatever Eichenwald and Shinefield had just done, it was apparently going to have a lasting effect. What was not yet known was whether the colonization by 502A would prevent colonization by 80/81.

Eichenwald and Shinefield were emboldened to take the next step. They found hospitals around the country in which 80/81 was both present and common. Or, rather, the hospitals found Eichenwald and Shinefield. The first was Cincinnati General, where Dr. James M. Sutherland, a neonatal specialist, was working. Sutherland called Eichenwald and Shinefield and asked for help. Sutherland's hospital had been plagued by 80/81 during the fall of 1961. Forty percent of newborns were colonized by this harmful bacterial strain. Soon, Shinefield was on the road, with a sample of Caroline Dittmar's 502A by his side, heading for Ohio. In Cincinnati, Shinefield and Sutherland inoculated the nostrils or umbilical stumps (or both) of half of the newborns in each of the nurseries at the hospital with 502A, the putative defender. The rest of the newborns were not inoculated. Which newborn was assigned to each treatment was random, as was the location where each newborn was placed among the hospital's three nurseries. Shinefield and his colleagues then examined whether the individuals inoculated with the putatively beneficial *Staphylococcus* stood a reduced risk of infection with 80/81. They were planting one species—a crop—and hoping that it would ward off another—a weed. They were farming. Like farmers, they hoped to reap what they had sown. They hoped what they had just planted was not a garden of terrible weeds (and infected babies).

The results of this study were important. They were important to each newborn infected with 80/81 or any other pathogen in each hospital around the world; they were important to the houses into which the newborns would move after leaving the hospital. They were probably of relevance to, at that point, hundreds of thousands if not millions of lives, particularly in the United States, where as many as twenty-five out of every thousand babies died in the hospital or soon after at home, most often because of infection.

Sutherland and Shinefield did not have to wait long for results. Of the babies inoculated with the potentially beneficial *Staphylococcus*, 502A, only 7 percent became colonized with 80/81, the pathogen. None of these cases in which the pathogen 80/81 colonized the baby happened in the hospital; all happened after the babies went home, presumably from 80/81 bacteria living somewhere undetected in their houses. That 502A was unable to ward off the pathogen in 7 percent of the cases was not ideal (the ideal, of course, being zero), but the key was how this compared to the babies who were not inoculated with the beneficial 502A. Babies who were not inoculated with the beneficial 502A were much more likely to be colonized by the pathogen 80/81—five times more likely. Sutherland's confidence in Eichenwald and Shinefield had been rewarded with concrete results.[16] Babies gardened with 502A, a strain of bacteria cultured off an individual nurse, Caroline Dittmar, could ward off 80/81, the dangerous weed, most of the time.

Shinefield was soon back on the road; Eichenwald didn't have the time to travel hospital to hospital, but Shinefield, as a new assistant professor, did. He would go on to repeat the study in Texas, where the results were similar, or perhaps even more promising. Just 4.3 percent of the babies inoculated with 502A were later colonized by the pathogen 80/81. In contrast, of the 143 infants in which the 502A was not seeded, 39.1 percent (nearly half) became infected with 80/81 or one of its close relatives. Just as in Cincinnati, the gardening seemed to work. Eichenwald and Shinefield would go on to repeat this experiment in Georgia (and write about it in a paper titled "The Georgia Epidemic") and then in Louisiana ("The Louisiana Epidemic").[17]

Gardening the body in self-defense seemed to unambiguously work. The strain 502A was an effective, safe defense against the most problematic pathogen in hospitals. But this was not enough.

Shinefield and Eichenwald would try something else. This is where events could have gone potentially very wrong. They noticed that after Shinefield's studies in Cincinnati and Texas, 80/81 entirely disappeared, briefly, from the nurseries. They decided to see whether they could use interference to more permanently eradicate 80/81 from hospitals.

Shinefield traveled hospital to hospital seeding newborns with *Staphylococcus aureus* 502A. He no longer used a control group. Now, he was just trying to cure children or, rather, to prevent them from ever suffering from infections in the first place. The results were astonishing. By 1971 four thousand newborns across the country had been successfully colonized, gardened as it were, with 502A. Not only did this reduce the prevalence of 80/81 in hospitals but also in some hospitals the team was able to eradicate the pathogen entirely. Gone. Done. On the basis of these results, Heinz F. Eichenwald concluded that "during the presence of a severe epidemic of staphylococcal disease, the use of 502A represents the most immediate, safest, and effective method of terminating the epidemic. I feel that we now have enough data, involving several thousand babies to indicate that this is a completely safe procedure."[18] Time would reveal just how the gardened beneficial strain *Staphylococcus aureus* 502A excludes pathogens such as *Staphylococcus aureus* 80/81. Beneficial strains of *Staphylococcus* produce enzymes that prevent the pathogens from forming biofilms; essentially, they prevent them from building their houses. They also produce bacteriocins toxic to other bacteria. The strain 502A uses bacteriocins to kill any species that tries to colonize where it has already become established.[19] Finally, 502A may also (inadvertently) trigger the immune system of the host in such a way as to make colonization of any additional bacteria less likely.[20]

In the immediate aftermath of this work, excitement boiled. The approach seemed like one that might spread ward to ward around the world. It would spread to homes, where people and surfaces could be inoculated. Doctors even began to inoculate adults with 502A, adults who were suffering from problems with infectious *Staphylococcus aureus*. With adults, the procedure was trickier. Doctors had to first use antibiotics to kill all of the pathogens in the nose (akin to weeding before planting), and then they could inoculate the adults with 502A just as they had the newborns. It worked 80 percent of the time. Shinefield, Eichenwald, and colleagues had, with 502A, invented an entirely new approach to medicine. What was more, the idea of interference had implications far beyond the establishment of single species of bacteria on the skin of newborn babies.

In 1959 the British ecologist Charles Elton published a book called *The Ecology of Invasions by Animals and Plants* in which he argued (among other things) that the more diverse a grassland, forest, or lake, the less likely it was that it would be invaded by newly introduced weeds, pests, and pathogens.[21] Evoking sentiments very similar to those of Shinefield and Eichenwald, Elton wrote that when animals invade diverse ecosystems they will "search for breeding sites and find them occupied, for food that other species are already eating, for cover that other animals are sheltering in, and they will bump into them and be bumped into—and often be bumped off." And the more diverse the ecosystem, the greater the odds that an invading species would be "bumped off." Elton also thought that the more diverse an ecosystem, the more likely it was that some predator or pathogen would be able to eat at and kill the invader. In general, Elton thought that more diverse ecosystems would be more resistant to invasion. Nearly sixty years of subsequent research have revealed that the pattern doesn't always hold—in ecology there are always caveats. Yet it often does. It holds often enough for ecologists, a people not ordinarily prone to hyperbole, to describe the ability of biodiverse ecosystems to resist invasion as a core part of the "planetary life support system."[22] A patch of goldenrod flowers growing in an old field is harder to invade if that patch contains more varieties of goldenrod,[23] or if it contains more kinds of herbivores. Elton's hypothesis was meant to explain patterns among plants and mammals. But it should work on bodies and homes too. Interference, then, on skin or elsewhere in our daily life, might work even better if two species, or even dozens, were to be intentionally grown. Imagine the students and grand-students of Shinefield and Eichenwald growing a diverse garden on your newborn, or on you, or in your bedroom.

Of course, what works among mammals and goldenrod stems might not work among microbes. The most elegant way to test Elton's hypothesis for microbes would be to construct microbial communities that differ in the number of species they contain. Such variation would mimic the natural differences in the diversity of microbial communities we see from one body to another, or one surface in a home to another. One could then introduce an invading species into such communities and see whether the invading species was less able to establish or persist in the more diverse communities. This didn't happen during Elton's lifetime (he died in 1991). But we can fast-forward a little and consider more recent research. Several years ago, a Dutch research group, led by the ecologist Jan Dirk van Elsas, carried out such a study. They did it in Petri dishes rather than on the skin of newborns, our perspectives on medical ethics having changed considerably since the 1960s.

Van Elsas and his colleagues filled flasks of sterilized soil with bacterial food and then populated those flasks with the same number of total cells of different numbers of strains of bacteria. The strains of bacteria were all isolated from the soil of grasslands in the Netherlands.[24] One treatment had five strains of bacteria, another twenty strains. Another still had a hundred strains. Then, as a final treatment, van Elsas and his team simply had the wild, ferocious diversity of real soil, with thousands of species therein. Control communities had no bacteria—just bacterial food. Van Elsas and his colleagues then introduced a nonpathogenic strain of *Escherichia coli* (aka the infamous *E. coli*) into each community and watched what happened over sixty days. Like 80/81, *E. coli* was the invader. The prediction was that the more diverse the community, the tougher it would be for *E. coli* to establish and persist. Competition for space, for key resources, and even for resources produced by other bacteria would exist. Also, the more diverse the community, the more likely that some of the bacterial strains would be able to produce antibiotics and, in doing so, kill whatever newcomers happened to show up before they ever had a chance. The niches of the community would be either fully occupied by competing bacteria or toxic.

When van Elsas and his colleagues grew *E. coli* on its own, it flourished, as one would expect to occur on a sterilized surface in your house later dusted with a little microbe food, be it cookies, dead skin, or whatever else. Over the sixty days of the experiment, the abundance of the *E. coli* remained steady and high. But when van Elsas added the *E. coli* to the soil in which five other strains of bacteria had been growing, *E. coli* grew more slowly and then began to disappear more quickly. When he added it to soil in which twenty or a hundred other strains were already growing, it disappeared more quickly still. When he added it to the wild diversity of the real soil, it became hard to even find the *E. coli* within the sample. The more diverse the bacteria, the harder it was for *E. coli* to thrive. In part, van Elsas would go on to show, this was because the numerous strains in the more diverse bacterial communities were using many kinds of resources more efficiently than were the fewer strains in less diverse communities.[25] Less was left over for *E. coli* to consume. The effect was even more extreme when van Elsas used an alternate approach, one that made the experiment an even better match for what might really happen in the soil. He created communities full of thousands of species of bacteria from the soil along with whatever bacteria-killing viruses were present as well.

The logical extrapolation of van Elsas's results to the conditions in our bodies or homes is to predict that pathogens are better at establishing on surfaces on our bodies or in our homes if these surfaces are less diverse and more sterile (and therefore the pathogens have less competition). The caveat is that this is true only if microbial food is available (it always is) and the house isn't totally devoid of life (but no house ever is). What a radical idea! Here was a possible extension of the approach used by Shinefield and

Eichenwald to the world around us. We could prevent the invasion of pathogens by favoring biodiversity on our bodies and in our homes. And the same should apply as well to insects (having a greater diversity of insects in your home, be they spiders, parasitoid wasps, or centipedes, should be more likely to keep pests such as house flies or German cockroaches in check). What is more, it would have the added benefit of increasing our exposure to the biodiversity of bacteria that, according to the biodiversity hypothesis, our immune systems need to function well. Here was a direct practical application of Elton's ecological insights.

IF SHINEFIELD AND EICHENWALD's ecologically minded "Eltonian" approach worked and spread hospital to hospital and even into homes, you might be wondering why it sounds so foreign. You might be wondering why you have never heard of babies or homes being gardened. You have never heard of any of this because beginning in the 1960s modern medicine decided to take a different road.

After its initial success, Eichenwald and Shinefield's idea enjoyed great popularity. It was to be the future! Then, it floundered. One fatality was associated with the accidental introduction of the "good" *Staphylococcus aureus* 502A into the blood of a newborn from a needle prick. Any bacteria able to make it into the bloodstream can cause an infection; once in the bloodstream the ordinary rules of good and bad, friend and foe, disappear. Also, a few of the inoculations of beneficial *Staphylococcus* led to skin infections (about one in a hundred). The infections were treatable with antibiotics, but they were infections all the same. The question was not whether these cases were a problem, it was whether they were a worse problem than what would have happened otherwise. They were not.

Eichenwald had noted early on that he and Shinefield had picked one of several possible approaches. One could garden beneficial strains that might interfere with and prevent the establishment of pathogens. One could also rewild the body and try to ensure it was colonized with a diversity of bacteria like that which might have covered our ancestors (minus the pathogens). Or one might "use various eradicative measures" that would kill *Staphylococcus* (or other pathogens) when infections occurred. One could garden, rewild, or kill. The third approach, Eichenwald went on to note, would have two problems. The pathogens would eventually evolve resistance to the eradicative measure leveled against them. Also, any attempt at eradication of the pathogen would kill both the good and the bad and, in the long term, make it easier for the bad to reinvade.[26] This is the situation we often face when we make a choice about how to manage the species around us.

In spite of Eichenwald and Shinefield's studies, collectively, hospitals, doctors, and patients chose the third approach, the killing. It seemed more sophisticated, part of a grand future in which we humans could control the world around us with ever more novel chemicals—be they antibiotics, pesticides, or herbicides. Though it had problems, we could figure them out in the future. The third approach also seemed, superficially, simpler. The antibiotic methicillin had become cheap and was easily available in hospitals. Using it was easy—nothing had to be grown, inoculated, or gardened. Methicillin was the first wave of a second generation of antibiotics, synthetic antibiotics engineered to be harder for the bacteria to evade. Methicillin was able to treat the *Staphylococcus aureus* 80/81 infections.

But even in those earliest days, it was recognized—and not just by Eichenwald and Shinefield—that eventually bacteria would adapt even to the new antibiotics, just as the pests and weeds would adapt to the pesticides and herbicides. Alexander Fleming, the discoverer of the antibiotic penicillin, had pointed out as much in his Nobel Prize speech in 1945.[27] It was widely recognized by scientists that using antibiotics, especially on newborns, tended to kill the pathogen but also favored a group of unusual bacteria that did not seem particularly likely to be beneficial. Shinefield had commented on as much in his work, but did so as if stating the obvious, something everyone should know. Thus, the early successes of using antibiotics were clear, but so too, to those who were paying attention, were the longer-term problems: antibiotics were easy to use, but they had negative side effects on other microbes, including those beneficially on and in our bodies, and would eventually become useless when resistance evolved. Resistance would take longer to evolve if the antibiotics were used in moderation when most necessary. Conversely, it would evolve more rapidly if antibiotics were used indiscriminately. In full awareness of each of these aspects of antibiotic use, the eradication approach was taken. Antibiotics were used frequently and largely without regard for whether they were absolutely necessary.

When Fleming and others predicted the evolution of resistance, they understood it was likely, but not how it would occur. We now understand well the ways in which bacteria adapt to antibiotics. In large populations of bacteria, some individuals are likely to have (or to develop) mutations that allow them to better survive antibiotics. Those bacteria need not be good competitors, just survivors, because the use of antibiotics kills the species with which they compete. The origin of such mutations and their increase in abundance in the presence of antibiotics can now be shown in the lab. In a recent experiment, for example, Michael Baym, Roy Kishony, and colleagues at Harvard Medical School set up a long tray (two feet by four feet) of bacterial food in a medium of agar. In this experiment, Baym, Kishony, and colleagues played a trick on the bacteria. They laced part of the agar in the rectangular Petri dish with antibiotics. On the left and right edges of the Petri dish, where Baym and his colleagues placed the bacteria, no antibiotics were present. But moving toward the center of the dish, the concentration of antibiotics increased until, at the very center of the Petri dish, it was far greater than that used clinically, a concentration in a microbial world that corresponds to the nuclear option in the larger world. The scientists then filmed what happened over time.

First, the bacterial strain grew over the agar where there were no antibiotics. It covered the agar completely—a lawn of life. But then the food in those areas became scarce. The bacteria stopped dividing. Just beyond this antibiotic-free zone lay a field of food, but it was laced with antibiotics. In this context, any bacteria with the ability to venture out to eat the food with antibiotics would be more likely to survive into the next generation and, having survived, to monopolize the food resources. They would do better even if they did not do as well as those bacteria that initially grew on the abundant food without antibiotics. None of the bacterial cells in the Petri dish at the beginning of the experiment had genes that would allow them to deal with antibiotics. Every single original bacterial cell was susceptible to these antibiotics. It could have stayed this way, and if it did, the bacteria would have stopped at the edge of the antibiotics and the experiment would have been over. But they did not stop.

In the short time in which the bacteria were growing, mutations emerged. Only a few emerged per generation. But the generations were so very fast in coming that soon there were bacteria that could grow on the low concentrations of antibiotics, a few strains that—via the dance of mutation, bacterial sex, and survival—had made it through. Quickly, they consumed food in the agar that had low concentrations of antibiotics, and, once more, they grew hungry. But, before long, a single bacterium developed a mutation that allowed it to colonize the agar filled with a higher concentration of antibiotics. Later, thanks to yet another mutation, the even higher-concentration agar was colonized as well, until the entire plate of agar was sucked of its food and covered in bacterial cells. All of this, the entire thing, an evolutionary masterwork of stunning genius and consequence, happened in eleven days. Eleven days.[28]

Eleven days seems quick, but it is slow compared to what happens in hospitals. In hospitals (and homes), bacteria don't have to wait for mutations. They can borrow genes from other bacteria that confer resistance to antibiotics. This is to say, in the real world all of this evolution would take far less time than eleven days. This is just what has happened again and again since the inoculation of babies with nonpathogenic bacteria was abandoned and the use of antibiotics became ever more common.

As a result of the overuse of antibiotics, the problems posed by resistant pathogens in hospitals are far worse than they were in the 1950s when 80/81 first appeared. They are worse not just among newborns but also more generally. Initially, some strains of 80/81 could be killed with penicillin (even if others couldn't). By the late 1960s, virtually all infections with *Staphylococcus aureus* were due to strains resistant to penicillin. Not long thereafter, some *Staphylococcus aureus* strains also evolved resistance to methicillin as well as to other antibiotics. By 1987, 20 percent of infections with *Staphylococcus aureus* in the United States were due to strains resistant to both penicillin and methicillin. By 1997, more than 50 percent were; by 2005, 60 percent. Not only is the proportion of infections due to resistant organisms increasing but so too is the total number of infections. As the proportion of infections caused by antibiotic-resistant bacteria has increased, so too has the number of antibiotics to which bacteria are resistant, both in the United States and globally. Many infections are now caused by *Staphylococcus* strains that are resistant to all but the antibiotics of last resort, antibiotics such as the carbapenems that doctors hold back in case things go really bad.[29] Some infections are now even resistant to the antibiotics of last resort. These infections cost the medical system billions of dollars a year in the United States alone and cause tens of thousands of human deaths per year.[30] Nor is the United States an exception—the trends are similar in much of the world. Nor is *Staphylococcus aureus* an exception. Resistance is also ever more common in the bacterium that causes tuberculosis (*Mycobacterium tuberculosis*) as well as bacteria associated with intestinal infections, such as *E. coli* and *Salmonella*. In some cases, the rising incidence of resistance results primarily from the overuse of antibiotics by humans. In others, it is due both to the overuse in humans and the use in domestic animals, where antibiotics are given to make pigs and cows get fatter quicker.[31]

Even in light of this inevitable evolution and increasing understanding of the consequences of the overuse of antibiotics, the response of many hospitals to the surge in resistant bacteria has been to step up the war on microbes, to run forward screaming a wild antimicrobial battle cry. Hand-washing regimens have been stepped up, which is a good decision, or at least not a bad one. Hand washing using soap, as far as we know, has no effect on the layer of normal bacteria on the skin, but it does wash away newly arrived species, which in hospitals are likely to be pathogens. But the proactive use of antibiotics has also increased in a no-holds-barred approach to fighting pathogens called "decolonization." In decolonization, the nasal passages of patients going into surgery, dialysis, or the intensive care unit are blasted with antibiotics to get rid of any *Staphylococcus aureus*. In the short term, this approach has been heralded by the hospitals that employ it.[32] In the long term, the consequences seem clear enough: decolonization will lead the noses of these patients to be colonized by hospital bacteria. It also favors more resistance. Medical history is being reenacted on the bodies of patients. But this time, something is different. Thanks to the ways in which we use antibiotics and fund research, the rate at which bacteria are evolving resistance to antibiotics is outpacing the discovery of new antibiotics, and this is unlikely to change.[33] Bacteria are evolving resistance to our antibiotics faster than we can replace the antibiotics. Yet the medical culture that has arisen since the work of Eichenwald and Shinefield sees little other way to control pathogens on bodies, in hospitals, or in homes. Nor is it alone. When it comes to the control in homes of insects, or fungi, the story is similar. We need another way.

It would be hard to restart Shinefield and Eichenwald's program. It would be harder to start a more ambitious effort to garden our whole homes or hospitals. Our perspective on risk has changed in ways that emphasize the dangers of gardening and largely ignore the dangers of war. This is bad news. But there is good news.

Antibiotic-resistant bacteria, like pesticide-resistant insects, are poor competitors. In the wild, most of these resistant organisms are weaklings. They are what ecologists call "ruderal species," species that survive only in environments where conditions are so chronically difficult that no other life-forms do well. Van Elsas showed that *E. coli* becomes less abundant and struggles to establish itself when soil microbial communities are diverse, but the *E. coli* that van Elsas studied was not resistant to antibiotics. If it were, we have every reason to believe that it would have struggled even more to persist in diverse communities. Like German cockroaches, antibiotic-resistant bacteria have fine-tuned their biology to the modern conditions we have created. They grow fast and take over our bodies and homes in the absence of competitors, viruses, and predators and in the presence of antibiotics. In such settings, resistant organisms thrive. But the compounds produced by the genes that confer resistance tend to be expensive for the bacteria—they require the bacteria to use the energy that they might have otherwise used to metabolize and divide. If there is no competition, living a slower, more expensive life doesn't matter. But where there is competition, this slower lifestyle puts resistant microbes at a disadvantage. This is one of the reasons that the very worst antibiotic-resistant bacteria are often confined to hospitals. In hospitals, they are constantly challenged by antibiotics. This quickly gets rid of any bacteria that aren't antibiotic resistant and rids the surviving bacteria of any competition. Even when the antibiotics are no longer being used, the competition is still held at bay, and so in hospitals, resistant microbes grow like nowhere else, released, like indoor German cockroaches, from competition, released and resistant to our assaults. Faced with competition, such species fail. Confronted with diversity, they fail. It is only in the unusual situations we have created on our bodies and in our homes that they succeed. This means that what we need to do to improve our situation may not necessarily be to fully garden the life around us; it may just be to rewild it a little. We need to find ways to tip the balance in our lives away from the pathogens and to let the biodiversity back in. We need to let biodiversity back in to help us fight deadly pathogens. We need to let biodiversity back in to help us fight chronic inflammatory disorders such as allergy and asthma. We need to let the biodiversity back in for these and many other reasons. And it can be simple, very simple. We've screwed up so badly that moderation can look like a panacea. We've screwed up so badly that inspiration for new ways forward may need to come from unlikely places and people, places such as kitchens, people such as bakers.

# 12

# THE FLAVOR OF BIODIVERSITY

I can't recommend anything about life indoors.

—JIM HARRISON, *A Really Big Lunch*

Tell me about a complicated man. Muse, tell me how he wandered and was lost.

—HOMER, *The Odyssey,* as translated by Emily Wilson

It is interesting to contemplate an entangled bank, clothed with many plants of many kinds, with birds singing on the bushes, with various insects flitting about, and with worms crawling through the damp earth, and to reflect that these elaborately constructed forms, so different from each other, and dependent on each other in so complex a manner, have all been produced by laws acting around us.

—CHARLES DARWIN, *On the Origin of Species*

ONE DAY WE HUMANS may garden exactly the species we need on our homes and in our bodies. We may be able to perfectly manage the species we most need in ways that yield a daily harvest that is, at once, healthful, beautiful, and sublime. Doing so will require extraordinary cleverness and a good understanding of the biology of most (if not all) of the species on our bodies and in our homes. I wouldn't hold my breath. This doesn't mean people won't soon be selling solutions in a bottle, jars of bacteria that you can spread around your home. They will. We just won't have very much understanding of whether such bacteria are actually beneficial. Instead of gardening, we need to rewild our homes; we need to let the wilderness back in, albeit a little selectively.

What I am advocating is not a return to a life in which we exert no control over the species that live with us. Instead, I am more humbly advocating moderation. We need drinking water in which the concentration of pathogens is low. We need effective hand washing to control the spread of pathogens person to person. We need everyone to be vaccinated for those pathogens for which vaccines exist. We also need the availability of antibiotics to treat bacterial infections when they arise. Nowhere is all of this clearer than in the many parts of the world where clean water, good hygiene and sanitation systems, vaccines, and antibiotics are lacking. But once and where we have done these things, once and where we have tamed the most dangerous beasts, we also need to find ways to allow the rest of biodiversity to flourish around us. We need, like Antony van Leeuwenhoek, to find joy and wonder in the bacteria, fungi, and insects in our daily lives.

If we do it right, inviting biodiversity back into our lives simultaneously helps to conserve biodiversity and allows us to avail ourselves of more of its services. The biodiversity of plants and soil can help our immune systems function properly. The biodiversity in our water systems can help keep pathogens in the water in check. If we pay attention to it, the biodiversity in and around our homes can help inspire wonder in children, as it did for Leeuwenhoek, as it does for me. The biodiversity of spiders, parasitoid wasps, and centipedes can help control pests. The biodiversity in our houses provides the opportunity, too, for discovery of enzymes, genes, and species useful to all of us, whether to make new kinds of beers or to transform waste into energy. Favoring biodiversity while at the same time dissuading dangerous species isn't rocket science and so is unlikely to ever be advocated by those who envy rockets. Instead, it is far more like making bread or kimchi, a reality I was reminded of recently when I sat down with Joe Kwon and Joe's mom, Soo Hee Kwon (aka Mama Kwon), to have lunch.

Joe, Soo Hee, and I had gotten together to talk about Korean cooking. Internationally, Joe is best known as the cellist for the popular band the Avett Brothers. The Avett Brothers play bluegrass-inspired rock, and Joe plays the low notes that hold the music up. But, at least in Raleigh, Joe is also known for his love of food. The unusual schedule created by touring with the band gives Joe long periods during which he can spend a day, say, roasting a pig. Joe and his pigs are well regarded enough that people search him out just to be able to sit with him during a long day of cooking. Roasting a pig well takes time, time enough to consider both the loveliness of pig and the grandeur of the universe.

But on this particular day, I was sitting with Joe not because of his music, or because of his cooking, but instead because of his mother's cooking. Joe's mother, Soo Hee, grew up in Korea, where she learned to cook traditional Korean dishes such as *Haemul Pajeon* (a kind of seafood pancake), *Jajangmyeon* (noodles in black bean sauce), and *Tteokbokk* (spicy stir-fried rice cakes). She learned the techniques necessary to make those dishes. She learned to make food that was, in and of itself, a kind of love. She made that food with her hands. Korean food is often made with heavy involvement of the hands. The hands roll cabbage. The hands cover fish in brine. The hands touch and maneuver each ingredient and do so with a subtlety and specificity that is both somehow deeply Korean and incredibly individual.

Making Korean food wouldn't have anything to do with houses but for one important concept, *sson mhat* (손 맛), where *sson* means "hand" and *mhat* means "taste." *Sson mhat* refers not to the food itself but instead to the flavor given to food by the person who makes it—literally by their hands, but figuratively by everything about who they are and how they touch, walk around, and work with food. Inspired by this idea, I wanted to explore a hypothesis with Joe and his mother, namely, that the microbes from the body of a Korean chef (traditionally a Korean woman) are precisely what gives her food a flavor different from that of her sister's or cousin's food.

Joe, Soo Hee, and I ordered some drinks and lunch. We started to eat and talk. I wanted to understand what Joe's mother thinks about *sson mhat* and what the word means to her. In Korean cooking, more than almost any other cuisine, foods are often fermented (meaning that sugars are chemically broken down by bacteria or fungi in a way that produces gas, acid, alcohol, or some mix thereof). The by-products of this fermentation add flavor and aromas to the food, such as the acidity and sour notes of yogurts. They make food intoxicating (when alcohol is the by-product). They also make the food toxic to other microbes. Alcohol kills most pathogens; so too, acidity. During the days when cholera was common in London, those who drank beer tended to be less likely to die of cholera than were those who drank water. The alcohol in the beer made that liquid safer to drink. Yogurt is safe to eat because the acidity in the yogurt prevents other microbes from colonizing. Acidity is measured on a scale from 0 to 14. Substances that have a pH of 7 are neutral, those with values higher than 7 are basic, and those with pH values lower than 7 are acidic. The pH of yogurt is typically around 4, similar to the pH of the stomach of a baboon.[1] The acidity of sourdough starters, kimchi, and sauerkraut is similar. The fermenting microbes that produce the acid (often species of the genus *Lactobacillus*) are tolerant of this acid, whereas most other species are not. Some fermented foods, such as Japanese *natto*, are alkaline, and this alkalinity has an effect similar to that of acidity. It keeps pathogens at bay. Species with the genes necessary to grow (typically slowly) in the presence of alcohol, high acidity, or high alkalinity almost never have the genes required of pathogens, genes that tend to entail fast growth. Fermentation, then, is not only a way of gardening species with favorable effects on our food; it is also a means of warding off pathogens. Fermented foods are ecosystems that weed themselves.

Because of the many benefits of fermentation, most human cultures have fermented foods. On my desk sits a compendium of thousands and thousands of different fermented foods of the world, most of them unstudied.[2] Some fermented foods, be they fermented shark or fermented seal stuffed with fermented auk, require a little getting used to. Many, though, are more familiar to the Western palate. Bread, vinegar, cheese, wine, beer, coffee, chocolate, and sauerkraut are all fermented. We eat fermented foods all the time, whether or not we realize we are doing so.

Among the most complex and biodiverse fermented foods is Korean kimchi. Kimchi is a Korean staple. The average South Korean eats eighty pounds of this good stuff a year. In making kimchi, the first step is to split, salt, and wilt cabbage. After several hours, the salt is washed off and the cabbage is further split or cut and mixed, by hand, with a pasty slurry of sweet rice, fish paste (itself fermented), shrimp paste (also previously fermented), ginger, garlic, and onions and radishes. The paste must be pushed into and around each cabbage leaf, with fingers and thumbs. It is massaged. It is worked. It is brushed and pushed anew. The result is then put into jars (sometimes small, more often enormous) and left to ferment. This is the basic plan, but the details vary greatly. Hundreds of kinds of kimchis are made, kimchis using different spices, different vegetables, and different steps. As it turns out, there may be as many different kinds of kimchi as there are people who make kimchi.

To my senses, kimchi is a delight. All humans have taste receptors for sweet, sour, salty, bitter, and umami. The umami taste receptor was discovered most recently (so you might not have learned about it in school). It detects flavors such as those found in some savory foods, including many meat dishes. The food additive MSG (monosodium glutamate) is so delicious because it tickles our umami taste receptor. Kimchi is one of the few vegetable-based foods that best satisfies the umami taste receptor (sun-dried tomatoes are another). I think of kimchi and joy, hand in hand; if I am eating kimchi, odds are that I am experiencing joy. But kimchi, Soo Hee told us, was not all joy when she was a little girl. It was hard work. The cabbage was ready in November. So too the radish that would be used along with the cabbage. The cabbage and radish needed to be harvested in huge quantities and then mixed in with the chili peppers and other ingredients. The kimchi made from the Napa cabbage and radish was important because it would be a key source of nutrients—the vegetable and protein that would accompany rice—for the entire winter. When Soo Hee was young, the winter in Korea was long and cold. Kimchi was delicious, but it was also part of survival, of getting through the winter. Kimchi, like other fermentations, was a means of storing food. It was a way to preserve vegetables so that they would last and last. And kimchi was also, as Joe's mom told me, among the foods with the strongest *sson mhat*. Each person's kimchi had a unique hand flavor.

Soo Hee sometimes leads kimchi cooking classes. In one of these classes, she said, she had cut up the ingredients for many people to work alongside her and make kimchi. They all made the kimchi. They did it in exactly the same way. They used the same ingredients. They followed the general motions of Joe's mom's hands. She was the instructor. She led, they copied. But the motions were not identical. So much about how a hand moves, about how it holds and works with a vegetable, is unique and gestural.

Weeks later, when the kimchis were all done, Soo Hee told me, each one tasted different. Each one, each person's kimchi, had a different hand flavor. Some were more sweet, others more sour. Some smelled a little fruity, others less. Some were delicious, others, well, Soo Hee said, less so. With this, I leaned in further. I ignored the food in front of me. I was becoming convinced that this hand flavor is due in part to the microbes on the bodies of the people making the kimchi, on their bodies and in their homes. The microbes in kimchi are of many species. Some of those species are likely to come from the cabbage itself, or from the radish. But they also include microbes known to be human bodily microbes. *Lactobacillus* species, for instance, are key to kimchi, as can be, even, *Staphylococcus.*[3] *Lactobacillus* species are common bodily microbes. Some species and strains are known to be gut microbes; others are vaginal. *Staphylococcus*, on the other hand, is a human skin microbe. Each one of these species and genera produces different enzymes, proteins, and flavors. Each one contributes something different to the final food.

When Joe's mom helped to make kimchi in the winter as a girl, the air was cold. The water in which the cabbage was soaked and wilted was also cold. Everything was cold. But making the kimchi was necessary. She labored over the giant buckets, again and again. It was certainly not, she conveyed, an unambiguous pleasure. And yet, it was part of who she was, this creation and fermentation.

Winter kimchi was one of many kinds of fermentation in Soo Hee Kwon's house when she was a little girl. Other kinds of vegetables were also fermented into summer kimchi. Crabs were fermented when they could be caught or afforded. Fish, too. If a food was not fermented in Joe's mom's house, it was fermented somewhere nearby. Soybeans were sometimes fermented with their own microbes into a paste (*Doenjang*) or a sauce (*Ganjang*) and in other cases with a special bacterium (*Chongkukjang*).[4] Red peppers were fermented, too, into a paste to be used as a flavoring (*Gochujang*). Fermented food could keep until the most desperate seasons. When such foods were fermented, the microbes from those foods must have spilled over onto each and every surface of the house. They must have risen up into the air. It is easy to imagine that the microbes of Joe's mom's house, the microbes of Joe's mom (and the rest of her family members), and the microbes of the foods themselves were part of one continuous story. Perhaps kimchi was flavored not just by some microbial hand taste but also by something for which there is not a Korean word, "house taste." And perhaps microbial hand taste and house taste come together to change the daily experience and well-being of everyone living in houses in which kimchi and other foods are regularly being fermented. I had been trying to find ways to favor a diversity of beneficial species in our homes and in and on our bodies, and here, in the form of kimchi, was one, maybe.

After talking to Joe Kwon and his mom, I wanted to launch a new project to understand the biology of hand taste, house taste, and whatever other sorts of tastes might be found. Kimchi is as good an example as we have for how the microbes around us and on us influence our food. But kimchi didn't seem like an ideal candidate for our first large-scale study of food. It is an acquired taste, one bound up in culture, history, and context. We could study instead some cheeses. Like kimchi, cheeses depend on many species. French Mimolette cheese, for example, depends on both bacteria from human bodies and those from cheese mites (*Tyrophagus putrescentiae*).[5] Or we could study the famous Sardinian cheese casu marzu made using body microbes and the wriggling, transparent larvae of household cheese flies (*Piophila casei*).[6] But these cheeses, like kimchi, are biologically very complex, foods for which more is understood by chefs and bakers than by scientists. They are also foods that don't appeal to everyone (casu marzu is actually illegal to produce and sell, though one can still find it). We needed to start with a food that was potentially interesting to bodily and household microbes, simple enough to be experimentally tractable, and appealing to nearly everyone. We needed to start with bread.

Leavened bread rises because microbes in the dough produce carbon dioxide that becomes trapped in air pockets in the bread. If you cut a loaf of leavened bread in half, each hole and opening is the result of the exhalation of a group of yeasts contained inside a kind of gluten dome. Without microbes, bread dough does not produce carbon dioxide. Without gluten, bread dough cannot catch the carbon dioxide produced by microbes. The very first breads were made with barley, which lacks enough gluten to make leavened bread, and so were unleavened.[7] By no later than 2000 BCE, Egyptian bakers had figured out how to make bread using emmer wheat. Emmer wheat contains gluten. Doughs made with emmer wheat can rise so long as the right microbes are present.[8]

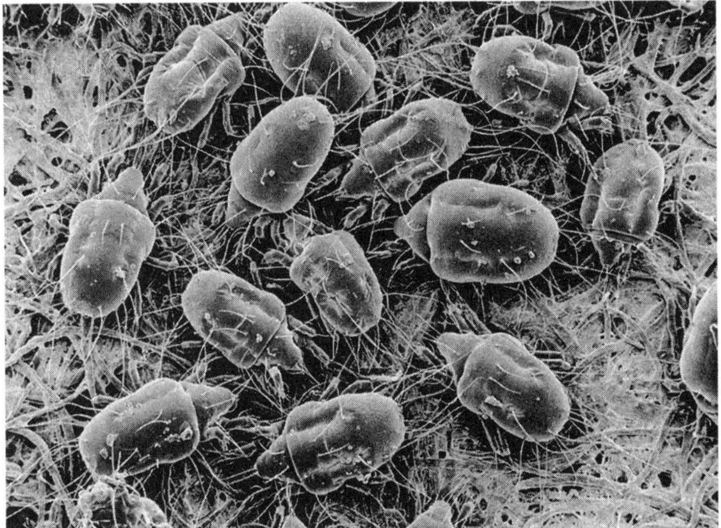

**Figure 12.1** Cheese mites happily doing their work as apprentices to the cheesemaker. *(Image by USDA Agricultural Research Service, USA.)*

The switch from unleavened to leavened bread can be seen in Egyptian art. Early Egyptian art shows flat loaves, but later loaves in similar scenes are round and risen. The microbes in those breads that enabled them to rise were yeasts. Yeasts in traditional bread produce carbon dioxide. Meanwhile, bacteria in those early breads would have made them taste sour. Nearly all traditional leavened bread is at least a little bit sour and this sourness (with very few exceptions) is often due to the same kinds of bacteria found in yogurt, species of *Lactobacillus*. We don't know how the ancient Egyptians controlled the yeasts and bacteria used in their breads,[9] but thanks to the depictions of risen bread in Egyptian art, we can be sure they did.

Today, the community of microbes used to make leavened bread is called a starter. To make a starter, one takes simple ingredients, often just flour and water, and leaves them out in a vessel.[10] Microbes ferment the starches in the flour.[11] Fed again and again with water and flour, the starter reaches a kind of steady state in which a relatively simple community of species survives in the bubbly, sticky, acidic mix. Just as with kombucha, sauerkraut, or kimchi, the more acidic the starter becomes, the less likely any pathogens are able to survive.[12] This is what we might hope for more generally in managing the life around us: simple ways of favoring species that benefit us and, at the same time, keeping troublesome species in check.[13] Starters, then, would be the ideal microbial communities for us to study; they are biologically diverse and, through their diversity, keep pathogens at bay.

A hundred years ago, essentially all leavened bread was made using a starter containing a mix of bacteria and yeast. Not anymore. In 1876, French scientist Louis Pasteur, progenitor of the germ theory (the idea that individual pathogen species can cause disease), discovered that some of the microbes that made beer and wine could also make bread rise. Not long thereafter, Emil Christian Hansen, a Danish fungus biologist, figured out that the key microbe in the fermentation of beer was a species of *Saccharomyces*. *Saccharomyces cerevisiae* was later shown to be sufficient to making a new kind of bread, one that was never sour, did not depend on any bacteria, and yet still rose. Scientists found ways to grow *Saccharomyces cerevisiae* in monoculture, in the lab, in enormous quantities, and then send it, freeze dried, around the world. The freeze-dried yeast allowed the production of bread to be scaled up. Today, the vast majority of bread you buy in the store is made using one of a handful of kinds of wheat and a single species of yeast, grown at large scales and then sold to the companies that make the bread.[14] This yeast goes by a variety of names, names that suggest diversity where there is none to be found. You don't have to be a nutritionist to know that, in the switch from homemade sourdough to a bag of mushy white bread, what has occurred is not exactly progress in terms of either nutrition or flavor. Industrial-scale breads don't have to be produced in this way, but most are. We have lost the richness of our daily bread, a richness of texture, flavor, nutrition, and microbes.

Fortunately, many home bakers and bakeries continue to make new starters as well as keep old ones alive. Much like their antecedents working a hundred or even a thousand years ago, these bakers mix flour and water, and then they wait.[15] In some cases, they repeat, exactly, what their ancestors did in making starters, step for step, gesture for gesture. In other cases, they make their own starters from instructions they found online. Either way, they must wait for microbes to begin to colonize the mix. They then take care of those microbes. The starters found in different bakeries and homes can be very different, but no one really knows why. More than sixty different lactic acid–producing bacterial species and a half dozen species of yeasts have been found in one or another starter. To understand why starters differ so much, we decided to conduct a study. The study would have two parts. In the first part, a true experiment, we would have each of fifteen bakers from fourteen countries make the same starter, using the same ingredients, where the only factors not controlled would be the bodies of the bakers and the air in their homes or bakeries. The bakers' bodies would be the treatment. We'd test the hypothesis inspired by my conversation with Soo Hee Kwon, namely, that the microbes on the bodies of the bakers and in their homes and bakeries would influence the microbes in the starters. In the second part of the study, a survey, we would characterize the microbes in starters from around the world.

For the first part of the study, the experiment, we teamed up with the Puratos Center for Bread Flavour in Saint Vith, Belgium. In the spring of 2017, Puratos helped us send out identical sourdough starter ingredients to each of the fifteen bakers in each of fourteen different countries. Each baker then mixed the flour and water and waited. Once the bakers had a living, functioning starter, they continued to feed it with flour we had sent. Later in the summer, we identified the microbes in each of these starters and whether they were contributed by the flour, the water, or the bakers' hands and homes. The "we" in this case was Anne Madden, an expert on the ecology and evolution of yeasts, and me.

Simultaneous with sending the starter ingredients to the bakers, we also began the second part of the project, a global survey of starters around the world. We invited people from Israel, Australia, Thailand, France, the United States, and anywhere else to share their starters with us. We reasoned that in the global sample we'd encounter new kinds of starter microbes, species present in just one region or even just one family. The experiment in Saint Vith enabled us to focus on how much starters varied when we held everything but the bakers constant. The global survey held nothing constant. In the global survey, we'd characterize the diversity of starters in all their glory. The participants in the global survey were people who, through making sourdough starters and bread, were helping to keep both tradition and microbes alive. They were curators of the beneficial biodiversity of bread microbes. The scientific team required to carry out the global part of the experiment would be huge and interdisciplinary. It would include, once more, Noah Fierer, but also Anne Madden, Liz Landis, Ben Wolfe, and Erin McKenney as food microbe experts, Lori Shapiro as an expert on the microbes of grains, with Angela Oliveira in charge of sequencing and analysis, Matthew Booker in charge of helping to record people's stories of food, Lea Shell and Lauren Nichols helping with everything, and many more people, not least of whom were the bakers who shared their sourdough starters. The home bakers and professional bakers who sent their sourdough guided every step of the work, more so than in any other project we have ever done.

In the global survey, when we talked to people about their starters, the number of questions we had grew and grew. Many of the starters had histories that were known to go back hundreds of years. Most starters had names. People talked about the starters like they were pets, but the attachment was even deeper. A mother could push her hands into the same starter her mother had cared for, which might be the same one her grandfather cared for or even her great-great-grandfather. And when people told stories about the starters, they did so as if describing a near immortal participant in their family's history. One starter, for instance, was called Herman. The woman who sent the starter called Herman included this note:

> In 1978, my parents went to Alaska. Because they knew I was a huge fan of sourdough, they brought back for me . . . [a] sourdough starter. This starter was over 100 years old. I rehydrated the starter, fed it, expanded it and began to use it. Because this starter was a living organism, we named him Herman and put him in our refrigerator, where he has lived for many years. We have used Herman to bake bread, rolls, waffles, etc. ever since. However, there is still more to my story. In 1994, two things happened that impacted our family. The first was the Northridge Earthquake, which caused a tremendous amount of damage in our area. The second was that just before the earthquake—and for the first time—Herman turned pink![16] That was disastrous because it indicated a bacteria had invaded our dear Herman and I had to toss him out. However, I was not overly concerned because my friend also had some Herman. Sometime after the earthquake, I finally got around to asking my friend for some Herman. When I did, her face dropped. It turned out that after the earthquake, as her husband was trying to clean things up, he noticed a jar of whitish grey somewhat sticky stuff in the back of the refrigerator. Thinking it was something old and bad—he threw it out! Disaster strikes again! My family was disconsolate. It was as if we had lost a well-loved family member. I tried buying and creating new starters, but they just did not have the same aroma or taste as Herman. In late 1993, my Mother had passed away. My Mom loved to entertain and, shortly before she died, she had been planning on having a party at their summer home. The following August, in 1994, my father, my siblings and I and our spouses decided to go to their summer home and give the party my Mom had been planning. When we got

> there, I realized that they had left precipitously when she was sick and that the refrigerator was in great need of cleaning. As I sat on the ground in front of the refrigerator sorting thru items, I began to laugh. And then cry. I knew as soon as I saw him, in all his gooey, sticky beauty. My Mom had a jar of Herman that I had given her at some point in time! Our kids doubted it could truly be Herman, but when we unscrewed the lid, Herman's pungent and unique aroma hit us right in the face. It was as though my Mom had reached down and given Herman back to us! Now, I have 4 jars of Herman. My kids and various friends also have him—just for insurance. I expect our story will continue to grow thru the generations of our family.

Participants, including Herman's owner, had questions. They wanted to know whether starters changed over time. They wanted to know whether their starter contained the same kinds of microbes it did a hundred years ago. They wanted to know whether the temperature at which the starter is kept really makes a difference. They wanted to know how to make starters that produced bread that was more sour or less sour.

In studying the starters from the global survey, we would try to answer as many of these questions as possible. We might be able to use the identity of the microbes present in these families of starters to trace their history genealogically (or, conversely, to find out that individual bacterial or yeast species die in or colonize starters so often that "Grandma's starter" no longer has very much to do with Grandma). We could and would try to identify the extent to which geography, climate, age, ingredients, and many other factors influenced which species were found in the starter. The microbes that colonize starters might be different in different regions. It was even possible that in some regions the local microbes might simply be unable to make starters. It had been speculated, for instance, that a traditional sourdough starter could not be made in the tropics, but no one seems to have ever studied whether this is true (no one except bakers in the tropics).

Meanwhile, we continued to obsess over the question we hoped to answer with the experiment in Saint Vith: Where do the sourdough microbes come from in the first place? To make sourdough, one mixes together flour and water, whether it's the cheap flour that comes in a paper bag at the store and tap water, or flour made from wheat hand-ground by a baker and mixed with the dew on dandelion leaves after the first full moon. Somehow, the right mix of bacteria and fungi appears. Poof!

In August 2017, the fifteen bakers, along with their fifteen experimental starters, all came to Saint Vith. Some bakers were younger, some were older. One worked at a bakery that supplies baguettes to thousands of stores a day. Another sold a few hundred loaves of bread a day, sometimes fewer, and made high-priced, well-known, and delicious toast. Some of the bakers used many starters while baking, each to suit a particular bread. Others used just one, a single starter to which they attributed a personality and gave a name. What all of the bakers shared was a deep, passionate, obsessive love for great bread. We met them all at the Puratos Center for Bread Flavour. The building was locked. The bakers gathered outside the center and waited to get in. The conversation was nervous and multilingual. It was nervous because the bakers were going to bake bread the next day, but they would do so using the experimental starters they had just made. These were not their ordinary starters. The bakers didn't want to bake bad bread. They didn't want to have made bad starters.

The door to the Center for Bread Flavour opened. We all walked in. After some introductions, Anne and I put the starters on the table and got ready to swab them. As we did, the bakers (who we imagined might just step back and watch) gathered around. They hunched in. They were used to being in control, used to being judged not for their starter but instead for what they could use a starter to make. The bakers wanted to care for their starters right away; they wanted to feed them.[17] They didn't want to wait. The bakers talked about what each of them thought would have been a better, more perfect way to make a starter. While these opinions were being voiced, Anne Madden put on gloves, I put on gloves and took out a notebook, and we began to sample. One by one, I opened the containers in which the starters were living. I inserted a cotton swab deep into each one, and then put the swab in a sterile case. Already, while carrying out this procedure, we could tell that the sourdoughs were different from each other. Some smelled extremely sour, others fruity, others a little bland. When Anne and I were done swabbing, we allowed the bakers to feed their starters. The bakers looked relieved; the starters, too. The starters bubbled gratefully and began, visibly, to rise.

The next morning, after the bakers had spent a night drinking Belgian beer (brewed by monks using a mix of bacteria and yeast) and singing songs about bread (really), and the starters spent a night luxuriating in a new dose of food, Anne and I came in to swab the bakers' hands. Anne did the swabbing. She took the work slowly, one hand at a time. She was careful to exhaustively sample all the cracks and crevices.

Finally, once each hand was swabbed, we let the bakers make dough with their starters. Each baker made the dough the same way. Or, rather, each baker made the dough according to the same written steps. So much about the relationship between baker and dough is unwritten and intimate that some of what happened next varied more from one baker to another than we would have liked. Some bakers were gentle with their dough, rolling it with a kind of tenderness. Others were aggressive. Some breads were coddled, others smacked. Some bakers used spoons, others wouldn't dare.[18] In the end, the experiment was also subject to the differences in the details of the traditions and styles of the bakers.

On the last night, Puratos hosted a bread and beer tasting. Each bread was set out. One by one, we sniffed the crust. We squeezed the bread and sniffed the insides, the crumb. We put the bread up to our ears and listened to the sound it made (or didn't make) when squeezed. We poked the bread to examine its elasticity. We chewed the bread on its own and then with a sip of beer. We savored the flavors of the slightly different microbes present in each loaf.

By this time, we had begun to believe that breads, like kimchi, are one way in which we experience the subtle biology of our homes. Our studies of homes and bodies have revealed the ways in which the microbes of each person and home are different. Such microbes must, we imagined, fall into starters. If and when they do, we taste bread and, whether we realize it or not, savor some of the species floating around us every day. Even species that can't be seen with the naked eye can be savored. In a single loaf of bread, glass of beer, or bite of kimchi or cheese, we find hints of the work the species around us do on our behalf. In French, the flavors associated with the soil, biodiversity, and history of a place are called *terroir*. When we bite or sip, we savor the *terroir*. Ecologists, more blandly, call the experiences that result from biodiversity the result of "ecosystem services." The ecosystem services of biodiversity in and around our home include the wonder biodiversity can inspire. They include the benefits biodiversity offers to our immune systems. They include the potential for new technologies, such as the use of camel cricket gut microbes to rid ourselves of industrial waste. They even include the local manifestations of distant services, such as the filtering of our tap water by the biodiversity in aquifers. I thought about this as we tried another bread, and another beer, and then another bread, and then yet another beer. I thought about this as we toasted "to bread" and "to microbes." I thought about it as I considered what the data from the Saint Vith study would show and the bakers started to sing again. "To bread, and to microbes!" And to a house in which both are delicious. "To bread and to microbes!" And houses in which we are all healthy. "To bread and to microbes!" And to lives filled with wild species we have yet to study or understand, species that float like mysteries all around us and offer services we are only beginning to measure. To bread and to microbes and our one wild life.

For a while, that was the end of the story of the Saint Vith experiment. The sourdoughs were made, breads baked, and samples shipped to the lab of my microbiological collaborator, Noah Fierer, at the University of Colorado, where their DNA would be sequenced, their species identified. In Colorado, the Saint Vith samples sat alongside the global samples. I thought this would be all I'd be able to say at the time of the publication of this book, all we would know. But just in case, I hounded Noah to hurry. Noah hounded Jessica Henley, his technician. Jessica hounded Angela Oliveira, a new student in Noah's lab. In December 2017, Angela sent us the results for both the Saint Vith and the global study. Usually, it takes months to fully make sense of results. But Anne Madden and I were so excited that we couldn't resist. We started to do analysis. I was in Germany. It was late at night. Anne Madden was in Boston. She still had a longer day ahead of her. We dug in.

When we had talked to the bakers about the Saint Vith project, we emphasized that the science that would be done on the samples of their starters would be hard. That wasn't totally accurate. It was better to say that parts of the Saint Vith experiment, like the global survey, could just fail. If they failed, we wouldn't have results we could believe and the whole effort would have been, though fun (really fun), scientifically useless. One way the project could have failed was if we didn't get enough DNA from the samples. There are a bunch of ways this might have happened. Fortunately, it didn't. Another way it could have failed was if the samples were contaminated, whether with microbes from my skin, microbes from Anne's skin, or even microbes that ended up inside the "sterile" swab containers when they were manufactured. But, when we checked our controls, we could see (and show) that we did not have contamination. There were even more boring ways this type of experiment could fail: a shipment could have failed to arrive (it happens all the time with scientific samples). The DNA could have degraded during shipping. Or an individual effort at sequencing the samples might have failed for reasons that were part bad magic, part technical, part human. None of these things happened either. The samples arrived. The box was not crushed. The samples were not spilled. The sequencing runs worked. We were able to process the data without trouble. We had, it seemed, the right mix of luck, hard work, and some more luck. But none of those things was what we were most worried about. What we were most worried about was that the results, particularly those of the Saint Vith study, would be inconclusive. This is what we didn't tell the bakers—that we might get the results but that we might not be able to tell whether their hands, lives, and bakeries had influenced their starters at all. It might simply be the case, even if the bakers' hands strongly influenced the starters, that amid all the other sources of variation, we wouldn't be able to say for sure. Fortunately, that isn't what happened.

As we started to consider the data, we discovered that the bacteria and fungi found in the Saint Vith starters were a subset of those we encountered in the global survey of starters. In the global survey, we found several hundred species of yeast and several hundred species of *Lactobacillus* and related bacteria. The starters weren't diverse compared to soil, homes, or even human skin, but they were more diverse than food scientists or bakers had previously understood. Different microbes were present in different regions. One fungus, for example, was confined nearly exclusively to Australia. Does it give Australian breads a unique taste? It might.

Among the starters made by the fifteen bakers who traveled to Saint Vith we found seventeen different species of yeast in the starters and twenty-two species of *Lactobacillus* bacteria. The diversity of bacteria and fungi in the Saint Vith starters was more or less what we might have expected given that we had sampled a relatively small number of starters and controlled the ingredients used to make them. We then looked at the results from the hands of the bakers.

On the basis of previous studies, we knew that all hands (just like noses, belly buttons, lungs, guts, and every other external surface of the body) are covered in a layer of microbes, a sheath. It is easy to imagine that in washing our hands, we remove all of the microbes. We don't. If you sample the microbes on someone's hands, have them wash and scrub their hands, and then sample the microbes again, no change in the overall composition of the microbes occurs. Noah Fierer was the first person to do a version of this experiment. The results were unambiguous and remain uncontested. Hand washing prevents the spread of pathogens and saves many lives a year, but it doesn't do so by sterilizing your hands. Instead, hand washing appears to remove microbes that have newly arrived, but not yet established on the hands. For example, when scientists experimentally put nonpathogenic *E. coli* on peoples' hands, washing with soap and water removed much of the *E. coli*. It didn't matter if the water was cold or hot. It didn't matter for how long people washed (so long as it was at least twenty seconds). Also, ordinary bar soap was more effective than antimicrobial soap at getting rid of the *E. coli*.[19] Keep washing your hands and do so with soap and water.

The most common microbes on hands in studies by Noah and researchers in other labs tended to be species of *Staphylococcus* (which dominates on the skin in general and is common in some cheeses, but not in bread), *Corynebacterium* (which cause armpit odors), and *Propionibacterium*.[20] *Lactobacillus* was also present on the hands. It was this *Lactobacillus*, along with its relatives, that we thought might be helping to inoculate sourdough. But *Lactobacillus* is usually relatively rare on hands—about 2 percent of the microbes on men and 6 percent on women in Noah's study.[21] Fungi can be present on hands, but are neither abundant nor diverse. This is what we expected on the bakers. We hadn't imagined there was any reason to expect differently. Hands are hands. Then, we looked at the results.

The first surprise was that the bakers' hands were totally different from any hands we had ever seen before. On average, 25 percent and up to 80 percent of all of the bacteria on the hands of the bakers were *Lactobacillus* and related species. Similarly, nearly all of the fungi on the bakers' hands were yeasts that can be found in sourdough starters, such as species of *Saccharomyces*. We had no idea this was even possible, and we don't yet fully understand it. My suspicion is that because the bakers spend so much time with their hands in flour (and starters), their hands become colonized by the bacteria and fungi they work around. One can even imagine a scenario wherein the *Lactobacillus* bacteria and *Saccharomyces* yeasts on bakers' hands outcompete other microbes by producing acid and alcohol, respectively. Such a community of microbes might make the bakers less likely to get sick than are other people. I'm speculating, but this result is really very novel and leads us down many new paths. I wonder whether all people who work with food develop unusual hand microbes. I wonder whether when more people cooked, a hundred years ago, or five thousand years ago, the continuity between food and hand microbes wasn't much greater in general than it is today. I wonder many things. We will have to do more experiments. And this wasn't the only exciting result.

When we looked at which bacteria were in which starters, we found that nearly all of the bacteria in the flour were also in the starters. No starter contained all of the bacteria from the flour, but most of the flour species were present in at least one of the starters. The species seeded into the starters by the flour included microbes from inside the grain seeds themselves that help the seed to grow (when the grain was milled, these microbes survived). They included soil microbes from wherever the grain was grown. But they were dominated by species able to live on the sugars of the grain and flour itself, including species of *Lactobacillus* bacteria. Results were similar for yeasts, with about half of the kinds of yeasts we found in starters coming from the flour. None of the bacteria nor the yeasts in the starters appeared to be those from the water. By now, we know the kinds of microbes that tend to be found in water, and they were absent from the growing starters. No *Delftia*, for example, the bacteria that can precipitate gold. Nor any *Mycobacterium*. The starters were not different because they used different sources of water. Why then were the different starters different?

In part, these differences were due to chance, which species from the flour happened to establish. In part, these differences were due to the bakers' hands. As we had hypothesized, the hands and lives of the bakers influenced the starters they made. The bacteria in each starter matched the bacteria from the hands of the baker who made the starter more than they matched the hands of other bakers. The same was true of the fungi, though to a lesser extent. The hands of the bakers were contributing bacteria and fungi (and, we presume, bacterial and fungal "hand flavor") to the starter. What is more, as we dug in to the details, the anecdotes were also telling. One of the bakers in our group is somewhat renowned for having a relatively unusual kind of fungus in his starter, *Wickerhamomyces*. That same baker had that same fungus in the starter he made in our experiment, and it was on his hands. His was the only starter with that fungus, and his hands were the only hands with that fungus. We also found yeasts and some bacteria in starters that didn't come from the flour, the water, or the bakers' hands—microbes that are most likely to have come from the life in the bakeries themselves.

When breads were baked using the starters, with identical ingredients (except for the microbes), the differences among starters influenced the flavor of the bread. Some starters made breads that were more sour, others that were more creamy, as judged by an expert panel of bread tasters. Each bread had a unique "microbe flavor," influenced by chance, and the microbes in the flour, on the bakers' hands, and in their bakeries. Once we consider in more detail the results and starters from our global survey, it is likely those starters, which are even more variable than those in the study of the bakers' breads, are able to create even more unique breads. Stay tuned. Meanwhile, everything we have learned so far suggests both that the species of microbes in a starter matter and that everyone is, to some extent, right about where the microbes are coming from. But we need to rethink this all a little bit. The way we initially asked our questions about the relationship between houses, bodies, and breads misses something important

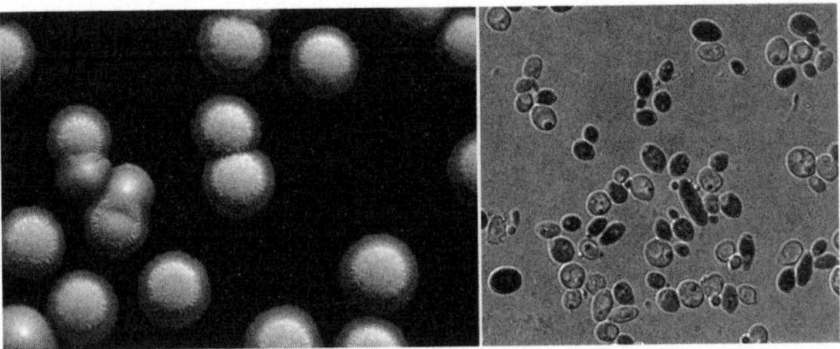

**Figure 12.2** Photos of colonies (left) and individual cells (right) of the yeast *Wickerhamomyces anomalus*. (Photographs by Elizabeth Landis.)

about what now seems to be going on, both with our food and our lives more generally. In bread making, the microbes on our bodies and in our houses are shaping the starter. But the starter is also shaping the microbes on our hands (and, potentially, in our homes). The action of making bread then is a kind of restoration, a restoration of certain kinds of biodiversity into our food, onto our bodies, and throughout our houses in such a way that all of these processes are connected. When we make sourdough starters, our bodies and homes flavor our daily bread. In making sourdough starters, the flour, starter, and bread enrich our bodies and homes. Nor should we imagine sourdough starters to be unique. The stories of cheeses, sauerkraut, kimchi, and many of the other foods we can ferment at home are likely to be similar.

At this point in our work, I estimate my colleagues and I have found roughly two hundred thousand species in homes. It is hard to accurately tally species from studies done at different times and with different methods (and the definition of a species depends on subfields, methods, and so on), but two hundred thousand is a reasonable estimate. Perhaps three-quarters of those have been bacteria found in dust, bodies, water, food, and guts. One-quarter is fungi. The arthropods, plants, and other taxa make up the rest. We haven't even started counting the viruses. But some houses are very diverse and other houses much less so, some houses full of species that seem mostly beneficial, other houses more likely to contain problem species. I imagined I'd get to the end of this book and conclude with the stories of architects, building engineers, and the like who had figured out how to build a healthy house full of the subset of these species that benefit us. I've spent thousands of hours researching this book. I did not find those people. Nor did I find their buildings. Sure, some new and innovative houses and cities do a better job than others of favoring biodiversity and beneficial species, but they do so not in their future-savvy sophistication but instead through their Paleo return to simplicity. They build houses with more open designs, out of more sustainable materials. This is great, but not yet a panacea.

I should have known at the beginning. One trouble with considering architecture as a solution is that much of what is offered to us by the most innovative architects is offered in small numbers, a house, a neighborhood, and at expensive prices. Such innovations are less likely to be offered to the big, collective "us." I won't be able to go and build a new house any day soon that perfectly favors biodiversity, much as I might want to. Also, the truth is, what people asked about when I told them about this book was not how they could build the perfect house. It was, instead, "Has studying the life in houses changed how you live?"

To that question there are some easy answers. I leave the windows open more often. I try to avoid turning on the central air-conditioning for as long as I can. If I've got the time, I wash dishes by hand to avoid spraying the fungus that lives in dishwashers everywhere around the house.[22] If water gets in my house, I get whatever got wet out. I considered getting a dog, but didn't (we travel too much). I begrudged my cat slightly more and spent a fair amount of time late at night wondering if she had given me *Toxoplasma gondii*. I planted a garden of fruit trees. I spent more time watching the insects in my home, and in other peoples' homes. I started sitting with my son to draw them, too, and of course wondered what new value each one might have (at the moment I'm pretty fascinated by the potential of silverfish). I started, too, to appreciate the magical services performed by the water from ancient, untreated aquifers. I savor the *terroir* of biologically diverse tap water. I buy more fresh food from local farmers, food that has some chance of still being covered in the microbes from the farm. All of those things. I didn't change my showerhead, but I do now eye the water coming out of it slightly suspiciously.

I was also inspired by the bakers. I started making more sourdough bread with my children. We also started experimenting more with different starters (I've got one going outside to see whether I can catch some interesting outdoor fungi with it). I was inspired by the lesson from the starter, namely, that there may be simple ways to favor beneficial biodiversity while keeping pathogens at bay, tricks of balance and moderation. That insight hasn't changed my life yet, but it changes how I think about my life. The biggest impact of the bakers came from the unexpected observation that the bakers' hands were covered in sourdough bacteria and fungi. The skin of the bakers reflects their daily actions. The truth is, all of our skin reflects our daily actions, so too do the species in our homes. In the Dark Ages, it was sometimes believed that God lived inside peoples' hearts and recorded on the heart's interior each good deed as well as each sin. The heart, we now know, is an unsentimental pump. But the biodiversity of your body and home is, indeed, a kind of record of your life, much as the bacteria on the bakers' hands are a measure of how much time they spend baking. I'll note here that once the bakers found out that some among them had hands covered in starter bacteria, each one wanted to know who had the most. Who, among them, had lived the life most fully submerged in bread?

This to me is the biggest lesson. The species in our homes are a measure of our lives. The early cave paintings of our ancestors documented the species they watched, stalked, and feared. The dust on our walls, in turn, documents the species with which we wake up each day. It is a measure of the species to which we are exposed and fail to be exposed. It is a measure of how we spend our moments. I know what I want my dust to say about me—that I am living a life embedded in biodiversity, a life in which I spend as much time outside with my family as I do indoors, a life exposed to biodiversity's grandeur and services, a life in which the species around me every day fill me with the sort of wonder the first microbiologist, Antony van Leeuwenhoek, felt. Leeuwenhoek woke up in his house each morning aware that most of life is benign or beneficial and that most of life, wherever you may be, remains to be studied. Leeuwenhoek lived at a time when the biodiversity around him was just beginning to be studied. So too do we.

# ACKNOWLEDGMENTS

When I read acknowledgments, I do so looking for secrets, a little bit about the book baker's magic. The first thing I can tell you about this book, if you are looking for secrets, is that more so even than the other books I have written it has emerged around the dinner table. Many of the stories in this book are the product of conversations with my wife, Monica Sanchez, and our kids about the life around us. So too the setting for so much of what is in this book is inspired by our time in our own house but also in houses around the world, the places we have stayed, and the many archaeological sites we have visited. In the interest of understanding the history of houses, our kids hiked up to ancient house sites in a dozen countries. They went through museum after museum looking at re-creations of ancient houses. They ran, with us, through the fields of Croatian farmers, searching for hidden Roman villas that have not yet been studied. They were lowered feet first into muddy caves to search for silverfish. They sat through all-day-long bread-making experiments, surrounded by bakers singing songs about bread. And, of course, they helped try out new projects, projects on backyard ants, basement camel crickets, sourdough microbes, and much more.

That, then, is the first secret, that my family helped make this book. The second secret is that so too did tens, maybe hundreds, of people I work with in my "lab" and companion labs at other institutions. I should explain that when scientists use the word *lab*, they sometimes mean just that—a laboratory space with high benches and the people who, like yet other pieces of furniture, occupy the lab. This isn't what ecologists tend to mean. Because much of what ecologists do is cheap and as likely to involve a bucket of mud as a fancy machine, to ecologists a lab is often a group of people who might share some physical space but who more often have fanned out across the world. My lab is a group of brains connected by a common set of quests. My lab is a group of people dedicated to beautiful new discoveries and to engaging the public in those discoveries. The work and thinking in my lab are connected to the work and thinking in other labs, be they in Colorado (Noah Fierer's lab), Massachusetts (Ben Wolfe's lab), San Francisco (Michelle Trautwein's lab), or any of a half dozen other places. People from this web of brains have contributed to every chapter. You met some of these people in the pages of this book, but just as many didn't appear. Many of those who are missing are missing in part because their participation is so central, so much a part of everything, that it is hard to describe just the role they play. This is a tricky element of science. We are asked all the time to say who did something, but we are terrible at really sorting it out.

Here are some examples of the people who helped make this book possible but who appear only fleetingly or not at all in these pages. Andrea Lucky and Jiri Hulcr came to my lab as a couple. They helped bring a new kind of community to my lab. Andrea launched our School of Ants project to engage the public in studying ants. Andrea and Jiri, along with an undergraduate student, Britne Hackett, launched our Belly Button Biodiversity project to sample belly buttons around the world to understand which skin microbes are common and which are rare (and why). During this same period, Meg Lowman joined the North Carolina Museum of Natural Sciences to lead the Nature Research Center. Meg had enthusiasm and cared deeply about engaging the public. She was key to the first steps we took in working with ants and belly buttons. The work with Meg and the museum was aided by Dan Solomon, then a dean in the College of Sciences, and Betsy Bennet, then the director of the North Carolina Museum of Natural Sciences, who together built a political and financial infrastructure that made it easy to work with the public on a truly grand scale. Little of the work we did on ants or belly buttons is really in this book, and yet it was this work that set the stage for much of what we would do in houses. It set the stage for this book.

Andrea and Jiri then left together for the University of Florida. Before they did, I hired Holly Menninger to run our projects engaging the public and undergraduate students in doing science. Holly was the one who figured out how we could actually organize projects to really reach people around the world and involve them in the science we were doing. Holly was also a voice of reason when I'd come into the lab with yet another crazy project and no new funds, time, or people to do it. Without Holly, little of the work we did on the biology of houses would have happened. She is now the director of public engagement and science learning at the Bell Museum in Minnesota, for which the museum and the whole damned state of Minnesota are fortunate. She doesn't occur much in the book because what she did was central to everything. Her work is always there—the social and intellectual infrastructure through which we figured out how to connect thousands of people in the process of doing science together.

With time, as Holly began to take on new roles (even before heading to Minnesota), including helping to orchestrate a cluster of Public Science at North Carolina State University (a group of new faculty all dedicated to engaging the public in science), Lauren Nichols and Lea Shell, along with Neil McCoy, took on more of the work of engaging the public in doing science. Lauren and Neil produced nearly all of the visuals in this book but also many of the other materials we use to talk about the life in homes. Lauren also helped to research this book, to follow loose threads as well as some threads that appeared well woven but that unraveled when she gave them a tug. Lauren read the book again and again and again. She formatted citations. She chased down leads. She helped rethink stubborn paragraphs and describe complex science. She responded to emails with titles like "Ahhh, copyedits are back and we have five days to go through the whole book. Can you drop what you are doing?" Thank you, Lauren. Lea Shell read the whole book and helped to make sure it included the things our participants most wanted to hear. Lea surveyed thousands of participants in our projects about the questions they wanted answered about the life in homes. Those answers are here, woven into this book. Hopefully, they are the questions you wanted answered, too.

In addition to my lab, the book depended on help from my collaborators, many of whom I now appear to owe favors. Noah Fierer you met in the book. Noah has been an outstanding collaborator, and I'm very grateful for the collaboration. He also read the whole book, thoughtfully, and then when I was worried about getting particular sections just right, he read them again. Carlos Goller has never formally been part of my lab but is often part of the most interesting science we do in the lab. Carlos has been an inspiration in terms of devising ways to engage students at the university in this work. Jonathan Eisen read the entire book and lent a critical eye to every bit and piece. Laura Martin helped me think about the history of human impacts on ecosystems. Catherine Cardelus, Katie Flynn, and Sean Menke offered thoughtful insights with regard to the ways the book might fit into university classrooms.

Many of the scientists featured in the book and those who work in fields related to the book helped with the book itself. They read chapters. They answered silly questions. Lesley Robertson welcomed me to Delft and spent two days talking and thinking about Leeuwenhoek and his work. Doug Andersen read the chapter on Leeuwenhoek and, like Lesley, helped me think through what he would have been like as a human. David Coil and Jenna Lang helped me understand the microbiology of the International Space Station. The chapter on showerheads was improved by comments from Matt Gerbert, a student in Noah's lab. I've never met Matt, but he does cool science. Jenn Honda helped me consider the medical microbiology of mycobacteria. Alexander Herbig and Johannes Krausse provided insights about the ancient human history of *Mycobacterium*. Christopher Lowry schooled me on the benefits of *Mycobacterium* species. Christian Griebler wowed me with the grandeur of aquifers and read the chapter on showerheads. Fernando Rosario-Ortiz read the same chapter and helped me think about water treatment.

Illka Hanski never saw this book, but emails with Illka helped me think about his work. Illka also read a previous version of the chapter that discusses his work. I only ever met Illka in person once, and it was when I was a graduate student. My lab mate Sacha Spector and I couldn't wait to talk to him about dung beetles and he didn't disappoint. I never imagined we'd reconnect many years later to think and talk about the life in homes. Niklas Wahlberg, one of Illka's former students, worked with me to get Illka's story right. Tal Haahtela and Leena von Hertzen helped me to understand their work and also put it in the context of the story of Karelia. Megan Thoemmes, Hjalmar Küehl, Fiona Stewart, and Alex Piel helped me think about the ecology of wild chimpanzees as it relates to our own ancestral ecology. Erin McKenney provided critical insights about food and feces, as she often does.

Nearly everyone who worked with me on the camel cricket project appears in the camel cricket chapter. They all read the chapter. Thank you, MJ Epps, Stephanie Mathews, and Amy Grunden. Jennifer Wernegreen has helped me, again and again, think about the evolution of bacteria associated with insects, as has Julie Urban. Genevieve von Petzinger helped me revisit the story of Paleo peoples in caves, as did John Hawks. The chapter on fungi was improved (repeatedly) by Birgitte Andersen who humors my musings about space stations and does work that many others find too hard to do. Birgitte also inspired me to think carefully about the basic biology of the fungi in homes. She reminded me, too, that even *Stachybotrys*, beast that it is, has a kind of beauty.

Martin Taubel helped me to think about the consequences of *Stachybotrys* in homes and what we do and don't know. Rachel Adams challenged me to think about how much we really know about what is and isn't alive and metabolizing in homes when it comes to fungi. It was Rachel who initially led me to consider the space stations.

The insect chapters were read and helped along by Matt Bertone, Eva Panagiatakopulu, Piotr Naskrecki, Allison Bain, Misha Leong, and Keith Bayless. Matt helped again and again. Thank you, Matt. Michelle Trautwein has been talking to me about this book for five years now, off and on, ever since we started working together in homes. Our work on the arthropods of homes, and conversations about arthropods and about life, all started when Michelle was still at the North Carolina Museum of Natural Sciences. It is my great good fortune that we have been able to continue them now that Michelle has moved to the California Academy of Sciences. Christine Hawn talked to me about the role of spiders in biological control. The cockroach chapter was improved by all of the entomologists in my ambit, among them Ed Vargo, Warren Booth, Coby Schal, Ayako Wada-Katsumata, and Jules Silverman, who dedicate all or parts of their research to figuring out how best to control the pests that even most entomologists don't like. Eleanor Spicer Rice (one of Jules's students) helped me think about how important the work on German cockroaches was to Jules Silverman. Thanks to the two department heads I had while writing this book (Derek Aday and Harry Daniels).

I started writing the chapter on Heinz Eichenwald more than five years ago. It wasn't quite right though. It wasn't until I joined a working group at the National Socio-Environmental Synthesis Center (SESYNC), led by Peter Jorgenson and Scott Carrol, that I really understood the extent to which Eichenwald's experiments offered up a path that we as a society decided not to take. Thanks to SESYNC and a huge thanks to Scott and especially Peter but also to the entire working group, including Didier Wernli. Thanks, too, to Kriti Shaarma, who thinks like a bacterium. Finally, thanks to Paul Planet for his insights but also for connecting me to Henry Shinefield. Henry was willing to share his story and help get the chapter just right. Henry remains a visionary, a kind one at that.

Jaroslav Flegr, Annamaria Talas, Tom Gilbert, Roland Kays, David Storch, Meredith Spence, Michael Reiskind, Kirsten Jensen, Richard Clopton, and Joanne Webster all read and helped with the chapter on dogs and cats. Thanks, too, to Meredith Spence for spending all of those years cataloging the parasites and pathogens of dogs (and to Nyeema Harris for inspiring Meredith's project). It has begun to pay off, Meredith! Nate Sanders, Neal Grantham, Brian Reich, Benoit Guenard, Mike Gavin, Jen Solomon, Joana Ricou, Annet Richer, and Anne Madden all helped with chapters of the book that were later deleted, chapters on forensics, wasps and yeasts, and the Pigeon Paradox. This book was once two hundred thousand words long, which is to say there is much more to the story of life in homes than I had space for here. Special thanks to the North Carolina State University Libraries and the wonderful people who work there. Karen Ciccone read the entire book and provided useful comments throughout. Mama Kwon, Joe Kwan, Josie Baker, Stefan Cappelle, Aspen Reese, Anne Madden, and Emily Meineke helped with the chapter on food. This book was winnowed, poked, and prodded by my agent, Victoria Pryor. Thank you, Tory. It was further subject to the supernatural selection of my editor, TJ Kelleher. TJ edited my first book, *Every Living Thing*. It is good to be, once again, working together. Huge thanks as well to Carrie Napolitano. TJ and Carrie, like many in the book world, always have too much to read and edit and too little time to do it, yet they managed to curate this work with great and perseverant care. The excellent copy editors Collin Tracy and Christina Palaia fixed broken sentences, mended problematic clauses, and more generally made sure each letter, comma, period, and colon was where it belonged. The book was supported with funds from the Sloan Foundation. Thank you, Sloan Foundation and, especially, Paula Olsiewski. The book was written while I was supported by an sDiv sabbatical fellowship and with help from and daily conversations with the scientists of the German Centre for Integrative Biodiversity Research (iDiv). Jon Chase, Nico Eisenhauer, Marten Winter, Stan Harpole, Tiffany Knight, Henrique Pereira, Aletta Bonn, Aurora Torres, and many others at iDiv helped me to revisit the biology of houses in light of the theory and insights of basic ecology.

Finally, I am immensely grateful to the many participants in our projects over the years. Many thousands of people have contributed to our projects studying homes. Those people opened their lives to our curious minds and joined us on a very strange quest. They asked questions that reframed our research. They inspired us and reminded us, again and again, of the joys of discovery and the even greater joys of discovery accompanied by multitudes. Thank you.

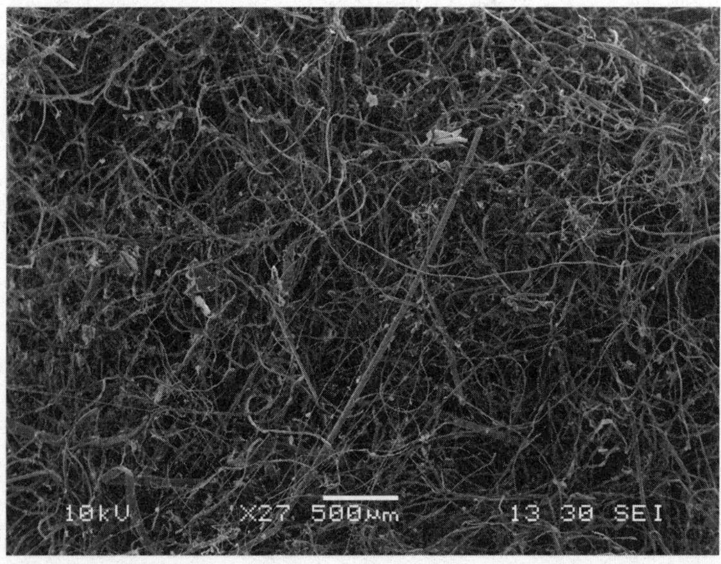


A microscopic image of dust. Dust is made up of many things, just as this book is a result of the influences of many, many people. *(Image by Anne A. Madden, with assistance from the Nanomaterials Characterization Facility of the University of Colorado, Boulder.)*

# NOTES

## PROLOGUE

1. N. E. Klepeis, W. C. Nelson, W. R. Ott, J. P. Robinson, A. M. Tsang, P. Switzer, J. V. Behar, S. C. Hern, and W. H. Engelmann, "The National Human Activity Pattern Survey (NHAPS): A Resource for Assessing Exposure to Environmental Pollutants," *Journal of Exposure Science and Environmental Epidemiology* 11, no. 3 (2001): 231. Or see, for example, results for Canada: C. J. Matz, D. M. Stieb, K. Davis, M. Egyed, A. Rose, B. Chou, and O. Brion, "Effects of Age, Season, Gender and Urban-Rural Status on Time-Activity: Canadian Human Activity Pattern Survey 2 (CHAPS 2)," *International Journal of Environmental Research and Public Health* 11, no. 2 (2014): 2108–2124.

## CHAPTER 1

1. Lesley Robertson, a microbiologist and historian, has been able to use microscopes like Leeuwenhoek's to see many of the kinds of organisms he would have seen, including diatoms, *Vorticella*, Cyanobacteria, and various species of bacteria. The work requires patience, wonder, and a willingness to try each and every permutation of lighting and specimen preparation, as did Leeuwenhoek himself. See L. A. Robertson, "Historical Microbiology: Is It Relevant in the 21st Century?" *FEMS Microbiology Letters* 362, no. 9 (2015): fnv057.

2. By the time Leeuwenhoek was using microscopes, most of his income probably came from a minor role he performed as a city official. That job offered Leeuwenhoek the leisure time affluence can afford, the sort of leisure that can feed obsession.

3. Leeuwenhoek would have used these lenses, called thread counters, to examine the quality of flax, wool, and textiles. See L. Robertson, J. Backer, C. Biemans, J. van Doorn, K. Krab, W. Reijnders, H. Smit, and P. Willemsen, *Antoni van Leeuwenhoek: Master of the Minuscule* (Boston: Brill, 2016).

4. The book is now available online for free through Project Gutenberg and contains wonders both very big and very small (https://www.gutenberg.org/files/15491/15491-h/15491-h.htm).

5. Samuel Pepys called it "the most ingenious book that I ever read in my life." See R. Hooke, *Micrographia: Or Some Physiological Descriptions of Minute Bodies Made by Magnifying Glasses with Questions and Inquiries Thereupon* (J. Martin and J. Allestrym, 1665).

6. At the time, it was not even believed that fleas reproduced; rather, people thought that fleas arose spontaneously from some perfect slurry of urine, dust, and their own feces. Leeuwenhoek documented the mating of the fleas (the smaller male dangling beneath the belly of the female). He documented the sperm and penis of the male (across his career he would document the sperm of more than thirty different animals, including his own). He found the eggs produced by the females. He sketched the eggs as they hatched, watched the larvae, and then saw their metamorphosis. He estimated that the process of sex, fertilization, eggs, and development could happen seven or eight times a year. He upped the ante, whether or not anyone happened to be paying attention. He did so all while carrying flea eggs with him wherever he went, in his bag, the way a child might carry a pet frog. See Robertson et al., *Antoni van Leeuwenhoek*.

7. De Graaf's accompanying letter can be read in full here: M. Leeuwenhoek, "A Specimen of Some Observations Made by Microscope, Contrived by M. Leeuwenhoek in Holland, Lately Communicated by Dr. Regnerus de Graaf," *Philosophical Transactions of the Royal Society* 8 (1673): 6037–6038.

8. Leeuwenhoek's timing was good. Science had begun its shift from a focus on revisiting old texts and abstracted thought to a focus on observation. Inspired by the work of the French philosopher René Descartes, this new generation of scientists believed that through observation one could most effectively discover new truths.

9. A. R. Hall, "The Leeuwenhoek Lecture, 1988, Antoni Van Leeuwenhoek 1632–1723," *Notes and Records the Royal Society Journal of the History of Science* 43, no. 2 (1989): 249–273.

10. Vacuoles are a remarkable storage device used by plant, animal, protist, fungal, and even bacterial cells. In vacuoles food can be stored. In vacuoles waste can be stored. In vacuoles conditions can be maintained that are different from those in the rest of the cell. In this way, vacuoles are perhaps most analogous to the clay vessels and reed baskets of early human civilizations;

vacuoles are multipurpose containers used by different species and at different times for different things.

11. The town in which Leeuwenhoek lived, Delft, was the epicenter of the study of the home, albeit by painters, not by scientists. The painters of Delft focused on depicting the townscape and then also the interiors of rooms. They depicted key habitats Leeuwenhoek would explore. Pieter de Hooch painted many scenes of courtyards; Carel Fabritius most famously painted *The Goldfinch* in its cage, but he also painted a landscape of Delft. Then there was Johannes (or Jan) Vermeer. Vermeer painted the same three rooms again and again, depicting small groups of people frozen, each set of them, in a kind of still life.

12. The lot where Leeuwenhoek's house once stood has never been excavated. It may contain lost microscopes, samples, or nearly anything else. This lot is now the site of a fancy coffee shop. Lesley Robertson and I tried to convince the owners to let us drill through their newly laid floor to search for artifacts of Leeuwenhoek's life beneath their shop. They declined, so instead I spent the next days trying to look through their windows into the backyard, the yard where Leeuwenhoek spent oh-so-very-much time.

## CHAPTER 2

1. The documentary, called *The Fifth Kingdom: How Fungi Made the World*, tells the story of fungi, their evolution, and its consequences. I was standing by the hot springs to speak about the evolution of fungi against a backdrop that was equal parts volcanism and microbes.

2. I suppose it is also possible that scientists can be frustrating! Though I think that the truth is the busy crew had their minds set on finding the perfect geyser and just forgot to count heads before driving off.

3. *Geyser* is actually an Icelandic word for "hot spring." For Brock's delightful autobiography, see: T. D. Brock, "The Road to Yellowstone—and Beyond," *Annual Review of Microbiology* 49 (1995): 1–28.

4. The archaea, like bacteria, evolved billions of years ago. Archaea are, like bacteria, single-celled. And, like bacteria, they lack a nucleus. The similarities end there, though. The cells of archaea are more different from those of bacteria than our cells are from plant cells. Archaea were discovered in the middle 1900s. Archaea are diverse but are often (though not exclusively) found in extreme habitats. They are never parasites of humans (ever). They are often relatively slow growing. And they have an extraordinary diversity of metabolic capabilities. I love bacteria, find them endlessly fascinating and surprising. But the archaea are even better, a life-form as old as life itself that never causes harm, carries out fundamental ecological processes, is poorly

studied, and, as we recently revealed, sometimes lives in places as immediate to your daily experience as your belly button. Leeuwenhoek missed them, which is to suggest that we are better than he was at navel gazing. J. Hulcr, A. M. Latimer, J. B. Henley, N. R. Rountree, N. Fierer, A. Lucky, M. D. Lowman, and R. R. Dunn, "A Jungle in There: Bacteria in Belly Buttons Are Highly Diverse, but Predictable," *PloS One* 7, no. 11 (2012): e47712.

5. Chemolithotrophs, chemical eaters that oxidize inorganic compounds to obtain energy.

6. Every species, be it a bacterium or a monkey, is given a species name and a genus name. The genus reflects the broader group to which a species belongs. As humans, we belong to the species *sapiens* (knowing) of the genus *Homo*. We are *Homo sapiens*. Just where the edges of one species end and another's begin is often fuzzy, even more so the boundaries of genera. In theory, one might argue that genera should tend to be named and grouped by scientists in such a way that a genus of primates and one of bacteria are about the same age. In practice, the ways in which scientists of different subfields decide how many species to include in a genus vary. The genera of bacteria tend to contain many species and be ancient (*Thermus* may well be tens of millions of years old, if not older). Genera of life-forms more like us tend to include fewer species and to be more recent; this difference is entirely a function of the preferences of microbiologists relative to, say, those of primatologists rather than differences between bacteria and primates themselves. The genus and species names of organisms are always italicized (as you will see in the text) unless a species has not yet been given a name, in which case the genus is italicized, but not the placeholder name of the species. For example, *Thermus* X1, where X1 means it is probably a new species but has not yet been given a species name. In most groups of organisms, apart from vertebrates and plants, many species bear these provisional names because no one has yet had a chance to name them formally even though their existence is known.

7. When Brock grew *Thermus aquaticus*, he was actually trying to grow a species he called simply "a pink bacteria" that lives in even hotter conditions. He was unable to grow the pink bacteria. Nor has anyone, it seems, been able to grow it since. For the first *Thermus* study, see T. D. Brock and H. Freeze, "*Thermus aquaticus* gen. n. and sp. n., a Nonsporulating Extreme Thermophile," *Journal of Bacteriology* 98, no. 1 (1969): 289–297.

8. R. F. Ramaley and J. Hixson, "Isolation of a Nonpigmented, Thermophilic Bacterium Similar to *Thermus aquaticus*," *Journal of Bacteriology* 103, no. 2 (1970): 527.

9. Economics would later borrow the term again, from ecology.

10. T. D. Boylen and K. L. Boylen, "Presence of Thermophilic Bacteria in Laundry and Domestic Hot-Water Heaters," *Applied Microbiology* 25, no. 1 (1973): 72–76.

11. J. K. Kristjánsson, S. Hjörleifsdóttir, V. Th. Marteinsson, and G. A. Alfredsson, "*Thermus scotoductus*, sp. nov., a Pigment-Producing Thermophilic Bacterium from Hot Tap Water in Iceland and Including *Thermus* sp. X-1," *Systematic and Applied Microbiology* 17, no. 1 (1994): 44–50.

12. Kristjánsson et al., "*Thermus scotoductus*, sp. nov.," 44–50.

13. One of the key points Brock makes, again and again, in his writing is that although industry has continued to work with the extreme microbes that he and his colleagues discovered in the 1970s and 1980s, very few researchers have continued to study the ecology of these organisms in the wild. See Brock, "The Road to Yellowstone," 1–28.

14. D. J. Opperman, L. A. Piater, and E. van Heerden, "A Novel Chromate Reductase from *Thermus scotoductus* SA-01 Related to Old Yellow Enzyme," *Journal of Bacteriology* 190, no. 8 (2008): 3076–3082. Also, because microbes never cease to surprise, another new strain of this same species has recently been shown to be able grow as a chemotroph when the need arises. In the lingo of scientists, it is a mixotroph. S. Skirnisdottir, G. O. Hreggvidsson, O. Holst, and J. K. Kristjansson, "Isolation and Characterization of a Mixotrophic Sulfur-Oxidizing *Thermus scotoductus*," *Extremophiles* 5, no. 1 (2001): 45–51.

15. For more on why so many bacteria are still unculturable, see S. Pande and C. Kost, "Bacterial Unculturability and the Formation of Intercellular Metabolic Networks," *Trends in Microbiology* 25, no. 5 (2017): 349–361.

16. "High throughput" is a fancy way of saying that one can do a lot at once, in this case decode the sequences of many organisms at the same time. It is "high-throughput" sequencing in much the way that McDonald's is high-throughput eating. And as for "next generation," such techniques advance so quickly that "next-generation" approaches are now feeling, among the hip and cool, oh-so "last generation," which was inevitable when the term was coined.

17. There are usually a few additional steps to help get rid of anything remaining in the sample that is not DNA. But this is the broad picture.

18. With time, the exploration spurred by the work of Brock and his colleagues and contemporaries led to the discovery of more thermophilic microbes—even hyperthermophilic microbes—and with them an entire library of their enzymes, each of which has slightly different abilities. A polymerase has been identified in *Pyrococcus furiosus*, for example, that works like Taq but that is even more stable at high temperatures.

19. The standard approach to sequencing does not identify organisms in such a way as to precisely match them up with named species. Instead, we get lists of life-forms grouped into their genera, *Thermus* 1, *Thermus* 2, and so on. Individual sequences are grouped into these names, these taxa, as a function of the similarity of their DNA sequences. Microbiologists call these taxa operational taxonomic units (OTUs) in recognition that they aren't quite species.

In some cases, a single OTU might really contain several species. In other cases, the reverse can be true (two OTUs belong in the same species). We are still at a phase in our naming of microbial life that all of this is a bit messy, so although OTUs are a highly imperfect way of grouping life, they provide a means of continuing to move forward while figuring out ways to reconcile new and old approaches to classifying life.

20. Recently, Regina Wilpiszeski used these techniques to search for additional thermophilic bacteria in hot water heaters, species in addition to *Thermus scotoductus*. When she did, she found a half dozen species of bacteria that are typically found only in hot springs, several of which are as yet uncultured and yet now nonetheless detectable.

## CHAPTER 3

1. More than once I walked long and far enough to exhaust all three, only to have to pick my way back to the station through the moonlight. In a forest chock-a-block full of venomous snakes, that was dumb.

2. S. H. Messier, "Ecology and Division of Labor in *Nasutitermes corniger:* The Effect of Environmental Variation on Caste Ratios" (PhD diss., University of Colorado, 1996).

3. B. Guénard and R. R. Dunn, "A New (Old), Invasive Ant in the Hardwood Forests of Eastern North America and Its Potentially Widespread Impacts," *PLoS One* 5, no. 7 (2010): e11614.

4. B. Guénard and J. Silverman, "Tandem Carrying, a New Foraging Strategy in Ants: Description, Function, and Adaptive Significance Relative to Other Described Foraging Strategies," *Naturwissenschaften* 98, no. 8 (2011): 651–659.

5. T. Yashiro, K. Matsuura, B. Guenard, M. Terayama, and R. R. Dunn, "On the Evolution of the Species Complex *Pachycondyla chinensis* (Hymenoptera: Formicidae: Ponerinae), Including the Origin of Its Invasive Form and Description of a New Species," *Zootaxa* 2685, no. 1 (2010): 39–50.

6. Only one paper has ever been written about this ant, in 1954. M. R. Smith and M. W. Wing, "Redescription of *Discothyrea testacea* Roger, a Little-Known North American Ant, with Notes on the Genus (Hymenoptera: Formicidae)," *Journal of the New York Entomological Society* 62, no. 2 (1954): 105–112. As for Katherine, I wasn't sure what she was up to, so I checked. She now works as a zookeeper at the El Paso Zoo. Katherine's interest in big cats ultimately proved more powerful than my power to distract.

7. This was work begun and led by Andrea Lucky, now an assistant professor at the University of Florida. A. Lucky, A. M. Savage, L. M. Nichols,

C. Castracani, L. Shell, D. A. Grasso, A. Mori, and R. R. Dunn, "Ecologists, Educators, and Writers Collaborate with the Public to Assess Backyard Diversity in the School of Ants Project," *Ecosphere* 5, no. 7 (2014): 1–23.

8. Long before we ever considered that one day we might be studying people's belly buttons or homes, Noah and I worked together on a project about ambrosia beetles, led by Jiri Hulcr. Jiri was studying the fungi and bacteria these beetles carry with them from place to place and garden to feed their babies. But this connection also allowed Noah and me to begin to work together. See J. Hulcr, N. R. Rountree, S. E. Diamond, L. L. Stelinski, N. Fierer, and R. R. Dunn, "Mycangia of Ambrosia Beetles Host Communities of Bacteria," *Microbial Ecology* 64, no. 3 (2012): 784–793.

9. Initially, these participants tended to be people we knew, but as our projects grew bigger, the scope of our engagement grew, and grew, and grew.

10. H. Holmes, *The Secret Life of Dust: From the Cosmos to the Kitchen Counter, the Big Consequences of Small Things* (Hoboken, NJ: Wiley, 2001).

11. Which meant Noah's technician, Jessica Henley, would soon be clipping four thousand tips of four thousand cotton swabs into four thousand vials. Sorry, Jessica. Sorry, and thank you.

12. In some places, the life in our homes records exactly where we put our bodies. Consider a study done by Matt Colloff, an ecologist and mite biologist then at the University of Glasgow. Colloff decided to sample and study, night after night, his own bed. He set up devices to monitor the temperature and humidity of nine quadrats of his bed while he slept. The bed was, Colloff notes in the study, a fifteen-year-old double bed with a fifteen-year-old mattress. The devices collected data about his mattress every hour on the hour as he snoozed. He expected to find more mites where conditions were warmer and more humid. This didn't seem to be the case. What he did discover was that wherever his body was, regardless of the temperature, there were also more mites. He found eighteen species of mites in total, including dust mites, but also predators of dust mites, all living beneath the spot on the bed where he slept, eating his body as it fell apart. One imagines that microbes, too, show a similar pattern, living in the greatest densities beneath those places we spend the most time. Colloff attributes the great diversity of his bed to the mattress's age. See M. J. Colloff, "Mite Ecology and Microclimate in My Bed," in *Mite Allergy: A Worldwide Problem*, ed. A. De Weck and A. Todt (Brussels: UCB Institute of Allergy, 1988), 51–54.

13. We later had a similar incident when studying the life in someone's belly button. The participant, a journalist of some renown, had a belly button filled almost exclusively with food-associated bacteria. We have no explanation. Some of life's mysteries are beyond the scope of science.

14. P. Zalar, M. Novak, G. S. De Hoog, and N. Gunde-Cimerman, "Dishwashers—a Man-Made Ecological Niche Accommodating Human Opportunistic Fungal Pathogens," *Fungal Biology* 115, no. 10 (2011): 997–1007.

15 Strain 121, as the species was called, was originally found near deep hydrothermal sea vents, where water can reach 130 degrees Celsius. It would turn out to survive at temperatures far beyond what anyone had previously believed possible. Autoclaves are like pressure cookers, pressurized so that they can reach sustained temperatures around 121 degrees Celsius (250 degrees Fahrenheit) and in doing so kill all life, especially the bacteria that contaminate lab equipment. Strain 121 could survive and thrive for more than twenty-four hours in the autoclave. Most autoclave sterilization cycles only last an hour or two. See K. Kashefi and D. R. Lovley, "Extending the Upper Temperature Limit for Life," *Science* 301, no. 5635 (2003): 934–934.

16. We later showed that this is less true for doors in apartments (which look like everything else in apartments). See R. R. Dunn, N. Fierer, J. B. Henley, J. W. Leff, and H. L. Menninger, "Home Life: Factors Structuring the Bacterial Diversity Found within and between Homes," *PLoS One* 8, no. 5 (2013): e64133.

17. B. Fruth and G. Hohmann, "Nest Building Behavior in the Great Apes: The Great Leap Forward?" *Great Ape Societies*, ed. W. C. McGrew, L. F. Marchant, and T. Nishida (New York: Cambridge University Press, 1996), 225; D. Prasetyo, M. Ancrenaz, H. C. Morrogh-Bernard, S. S. Utami Atmoko, S. A. Wich, and C. P. van Schaik, "Nest Building in Orangutans," *Orangutans: Geographical Variation in Behavioral Ecology*, ed. S. A. Wich, S. U. Atmoko, T. M. Setia, and C. P. van Schaik (Oxford: Oxford University Press, 2009), 269–277.

18. Three-toed sloths make the treacherous descent from the safety of their canopy perches down to the forest floor every three weeks or so to defecate. When they do, the moths living in their fur lay their eggs in the sloth dung. The moth larvae develop completely within the dung. When they are mature, they fly up to the canopy to take up residency in the sloth's fur. An individual three-toed sloth can harbor between four and thirty-five moths. It has been suggested that the moths provide nutrients that help algae, also growing in the sloth's fur, to flourish. The sloths then eat the algae to supplement their diet because the algae are richer in lipids than is the foliage. See J. N. Pauli, J. E. Mendoza, S. A. Steffan, C. C. Carey, P. J. Weimer, and M. Z. Peery, "A Syndrome of Mutualism Reinforces the Lifestyle of a Sloth," *Proceedings of the Royal Society B* 281, no. 1778 (2014): 20133006.

19. See, for example, M. J. Colloff, "Mites from House Dust in Glasgow," *Medical and Veterinary Entomology* 1, no. 2 (1987): 163–168.

20. The chimpanzees don't go to the bathroom in their nests, they don't appear to abandon much of their food, and they build a new nest most nights.

All of this must help to keep the microbes and other life-forms associated with chimpanzee bodies from accumulating. See D. R. Samson, M. P. Muehlenbein, and K. D. Hunt, "Do Chimpanzees (*Pan troglodytes schweinfurthii*) Exhibit Sleep Related Behaviors That Minimize Exposure to Parasitic Arthropods? A Preliminary Report on the Possible Anti-vector Function of Chimpanzee Sleeping Platforms," *Primates* 54, no. 1 (2013): 73–80. For Megan's study, see M. S. Thoemmes, F. A. Stewart, R. A. Hernandez-Aguilar, M. Bertone, D. A. Baltzegar, K. P. Cole, N. Cohen, A. K. Piel, and R. R. Dunn, "Ecology of Sleeping: The Microbial and Arthropod Associates of Chimpanzee Beds," *Royal Society Open Science* 5 (2018): 180382. doi:10.1098/rsos.180382.

21. H. De Lumley, "A Paleolithic Camp at Nice," *Scientific American* 220, no. 5 (1969): 42–51.

22. It is hard to imagine that hominids moved into Europe more than 1.7 million years ago without the ability to build a shelter. The trouble is that the elements out of which the first homes would have been built—branches, leaves, and mud—don't preserve well. But it doesn't take many steps to go from building a nest, to building a wind shelter, to building a crude dome.

23. L. Wadley, C. Sievers, M. Bamford, P. Goldberg, F. Berna, and C. Miller, "Middle Stone Age Bedding Construction and Settlement Patterns at Sibudu, South Africa," *Science* 334, no. 6061 (2011): 1388–1391.

24. J. F. Ruiz-Calderon, H. Cavallin, S. J. Song, A. Novoselac, L. R. Pericchi, J. N. Hernandez, Rafael Rios, et al., "Walls Talk: Microbial Biogeography of Homes Spanning Urbanization," *Science Advances* 2, no. 2 (2016): e1501061.

25. We humans tend to kill off useful species in our houses and, at the same time, accidentally favor the bad ones. The termites in our homes do the opposite. Formosan termites (*Coptotermes* spp.), for example, can smell fungi on their bodies or in their nest by waving their antennae in the darkness of their chambers. They are also then able to clean individual fungal spores off of their bodies. Once fungal spores are encountered, the termites get rid of them by eating them. Termite guts effectively encapsulate the fungi in feces, which serves as an effective biocide, much as does the nacre of an oyster's pearl around the tapeworm cyst it includes. The termites then build the walls of their nests with their feces, antimicrobial spit (their own), and soil. Inside those walls the fungi live on, trapped. By employing this collection of behaviors—detection, consumption, construction—these termites have created an environment that is largely devoid of their most serious foes while, at the same time, letting other species, including those they depend on for digestion, persist unharmed. See A. Yanagawa, F. Yokohari, and S. Shimizu, "Defense Mechanism of the Termite, *Coptotermes formosanus* Shiraki, to Entomopathogenic

Fungi," *Journal of Invertebrate Pathology* 97, no. 2 (2010): 165–170. Also see A. Yanagawa, F. Yokohari, and S. Shimizu, "Influence of Fungal Odor on Grooming Behavior of the Termite, *Coptotermes formosanus*," *Journal of Insect Science* 10, no. 1 (2010): 141. Also see A. Yanagawa, N. Fujiwara-Tsujii, T. Akino, T. Yoshimura, T. Yanagawa, and S. Shimizu, "Musty Odor of Entomopathogens Enhances Disease-Prevention Behaviors in the Termite *Coptotermes formosanus*," *Journal of Invertebrate Pathology* 108, no. 1 (2011): 1–6.

26. D. L. Pierson, "Microbial Contamination of Spacecraft," *Gravitational and Space Research* 14, no. 2 (2007): 1–6.

27. For bacteria. We will return to the fungi. See Novikova, "Review of the Knowledge of Microbial Contamination," 127–132. Also see N. Novikova, P. De Boever, S. Poddubko, E. Deshevaya, N. Polikarpov, N. Rakova, I. Coninx, and M. Mergeay, "Survey of Environmental Biocontamination on Board the International Space Station," *Research in Microbiology* 157, no. 1 (2006): 5–12.

28. The longest-term study found dozens of genera of bacteria, the most common of which were armpit bacteria (*Corynebacterium*) and acne bacteria (*Propionibacterium*). See A. Checinska, A. J. Probst, P. Vaishampayan, J. R. White, D. Kumar, V. G. Stepanov, G. R. Fox, H. R. Nilsson, D. L. Pierson, J. Perry, and K. Venkateswaran, "Microbiomes of the Dust Particles Collected from the International Space Station and Spacecraft Assembly Facilities," *Microbiome* 3, no. 1 (2015): 50.

29. S. Kelly, *Endurance: A Year in Space, a Lifetime of Discovery* (New York: Knopf, 2017), 387.

## CHAPTER 4

1. First discussed in a paper by Ron Pulliam. See H. R. Pulliam, "Sources, Sinks, and Population Regulation," *American Naturalist* 132 (1988): 652–661.

2. Dan Janzen has suggested some bacteria produce repulsive odors not as waste products but instead as a means of preventing their food from being eaten by us. They stink, he argues, in order to be able to eat in peace. Sometimes I have the idea that people next to me on planes are trying the same strategy. See D. H. Janzen, "Why Fruits Rot, Seeds Mold, and Meat Spoils," *American Naturalist* 111, no. 980 (1977): 691–713.

3. Just which odors we perceive to be disgusting is a reflection of both our evolutionary past and our culture. Culture modulates the way in which we think about a particular odor (for example, how we feel about fish paste). Evolution, however, shaped whether the signals in our brain triggered by an odor are perceived as unpleasant. It is worth noting that these perceptions are always species specific. The same "miasmic" odors that repulse us trigger the opposite reaction in a dung beetle or a turkey vulture.

4. This wasn't technically the story of the biology of a home, but when everyone gets their water from a common well in the city, the whole city and its biology spill over into the home.

5. One reason some cholera epidemics subside is because viruses (vibriophages) attack *Vibrio cholerae*. As *Vibrio cholerae* becomes abundant, so too do the vibriophages until they are so abundant that populations of *Vibrio cholerae* crash. Then populations of the vibriophages crash, which allows *Vibrio cholerae* populations to grow again. In the Ganges, the rise and fall of *Vibrio cholerae* and its virus are seasonal as are cholera cases. S. Mookerjee, A. Jaiswal, P. Batabyal, M. H. Einsporn, R. J. Lara, B. Sarkar, S. B. Neogi, and A. Palit, "Seasonal Dynamics of *Vibrio cholerae* and Its Phages in Riverine Ecosystem of Gangetic West Bengal: Cholera Paradigm," *Environmental Monitoring and Assessment* 186, no. 10 (2014): 6241–6250.

6. Inasmuch as millions of people still die every year from cholera, the challenge is to ensure such systems are available to everyone. The challenge is no longer figuring out the cause of the disease or even how to stop it but figuring out how to get the solution, clean drinking water, to everyone in the world. The challenge is no longer prevention of a mystery ailment due to miasma but instead the hard-to-resolve dilemma of global inequality and geopolitics.

7. I. Hanski, *Messages from Islands: A Global Biodiversity Tour* (Chicago: University of Chicago Press, 2016).

8. As if a kind of premonition, Haahtela referenced just twenty-three papers in this article, two of them by Hanski. See T. Haahtela, "Allergy Is Rare Where Butterflies Flourish in a Biodiverse Environment," *Allergy* 64, no. 12 (2009): 1799–1803.

9. United Nations, *World Urbanization Prospects: The 2014 Revision. Highlights* (New York: United Nations, 2014), https://esa.un.org/unpd/wup/publications/files/wup2014-highlights.pdf.

10. E. O. Wilson, *Biophilia* (Cambridge, MA: Harvard University Press, 1984).

11. See, for example, citations and discussion in M. R. Marselle, K. N. Irvine, A. Lorenzo-Arribas, and S. L. Warber, "Does Perceived Restorativeness Mediate the Effects of Perceived Biodiversity and Perceived Naturalness on Emotional Well-Being Following Group Walks in Nature?" *Journal of Environmental Psychology* 46 (2016): 217–232.

12. R. Louv, *Last Child in the Woods: Saving Our Children from Nature-Deficit Disorder* (Chapel Hill, NC: Algonquin Books, 2008).

13. D. P. Strachan, "Hay Fever, Hygiene, and Household Size," *BMJ* 299, no. 6710 (1989): 1259.

14. L. Ruokolainen, L. Paalanen, A. Karkman, T. Laatikainen, L. Hertzen, T. Vlasoff, O. Markelova, et al., "Significant Disparities in Allergy Prevalence

and Microbiota between the Young People in Finnish and Russian Karelia," *Clinical and Experimental Allergy* 47, no. 5 (2017): 665–674.

15. L. von Hertzen, I. Hanski, and T. Haahtela, "Natural Immunity," *EMBO Reports* 12, no. 11 (2011): 1089–1093.

16. This project would ultimately, despite being started amid favorable conditions, fall apart. Janzen was left to lead it with little in the way of funds and based on field work and taxonomy done by a handful of dedicated friends. See J. Kaiser, "Unique, All-Taxa Survey in Costa Rica 'Self-Destructs,'" *Science* 276, no. 5314 (1997): 893. Needless to say, it isn't done. It may well never be done.

17. The same effort in Raleigh, for instance, would be an incredibly onerous undertaking involving many hundreds, perhaps thousands, of multicellular species, to say nothing of the bacteria.

18. I. Hanski, L. von Hertzen, N. Fyhrquist, K. Koskinen, K. Torppa, T. Laatikainen, P. Karisola, et al., "Environmental Biodiversity, Human Microbiota, and Allergy Are Interrelated," *Proceedings of the National Academy of Sciences* 109, no. 21 (2012): 8334–8339.

19. H. F. Retailliau, A. W. Hightower, R. E. Dixon, and J. R. Allen. "*Acinetobacter calcoaceticus:* A Nosocomial Pathogen with an Unusual Seasonal Pattern," *Journal of Infectious Diseases* 139, no. 3 (1979): 371–375.

20. N. Fyhrquist, L. Ruokolainen, A. Suomalainen, S. Lehtimäki, V. Veckman, J. Vendelin, P. Karisola, et al., "*Acinetobacter* Species in the Skin Microbiota Protect against Allergic Sensitization and Inflammation," *Journal of Allergy and Clinical Immunology* 134, no. 6 (2014): 1301–1309.

21. Fyhrquist et al., "*Acinetobacter* Species in the Skin Microbiota," 1301–1309.

22. Ruokolainen et al., "Significant Disparities in Allergy Prevalence and Microbiota," 665–674.

23. Fyhrquist et al., "*Acinetobacter* Species in the Skin Microbiota," 1301–1309.

24. L. von Hertzen, "Plant Microbiota: Implications for Human Health," *British Journal of Nutrition* 114, no. 9 (2015): 1531–1532.

25. We understand so little that the answer may yet be very complex. For example, Megan has also explored the prevalence of Gammaproteobacteria in traditional Himba houses in Namibia compared to houses in the United States. Hanski and colleagues predict that we should see more Gammaproteobacteria in the Himba houses, houses made of mud and dung, houses out in the bush, than in houses in the United States. Megan sees the reverse. If this stuff were easy, we would have already figured it out.

26. M. M. Stein, C. L. Hrusch, J. Gozdz, C. Igartua, V. Pivniouk, S. E. Murray, J. G. Ledford, et al., "Innate Immunity and Asthma Risk in Amish and Hutterite Farm Children," *New England Journal of Medicine* 375, no. 5 (2016): 411–421.

27. T. Haahtela, T. Laatikainen, H. Alenius, P. Auvinen, N. Fyhrquist, I. Hanski, L. Hertzen, et al., "Hunt for the Origin of Allergy—Comparing the Finnish and Russian Karelia," *Clinical and Experimental Allergy* 45, no. 5 (2015): 891–901.

## CHAPTER 5

1. J. Leja, "Rembrandt's 'Woman Bathing in a Stream,'" *Simiolus: Netherlands Quarterly for the History of Art* 24, no. 4 (1996): 321–327.

2. Although Noah and I had both forgotten, this would actually prove to be (a check through my email reveals) the second time we talked about doing a project on showerheads. The first time the project went nowhere. The email chain died. This email from Noah was then a bit of a resurrection of an earlier enthusiasm.

3. A more complete list of the invertebrates in Danish water includes seed shrimp, flatworms, *Cyclops* species, species of *Tubifex,* bristle worms, amphipods, and roundworms. See S. C. B. Christensen, "*Asellus aquaticus* and Other Invertebrates in Drinking Water Distribution Systems" (PhD diss., Technical University of Denmark, 2011). See also S. C. B. Christensen, E. Nissen, E. Arvin, and H. J. Albrechtsen, "Distribution of *Asellus aquaticus* and Microinvertebrates in a Non-chlorinated Drinking Water Supply System—Effects of Pipe Material and Sedimentation," *Water Research* 45, no. 10 (2011): 3215–3224.

4. We know this thanks to work by Carlos Goller and North Carolina State University students. Carlos is now busy searching, one spigot to the next, for new varieties of these unusual bacteria. He has solicited the help of thousands of undergrads in this effort, thousands of undergrads whom he has asked to peer up into their faucets, searching for new life-forms. They have found not only *Delftia acidovorans* but also many other *Delftia* species, quite a few of which appear to be new to science.

5. Much as with the plaque on your teeth.

6. Biofilms allow microbes to hold fast, and they protect microbes from everyday dangers, such as those posed by humans. The concentration of antimicrobials required to kill bacteria in biofilms, for instance, is up to a thousand times greater than that required when they are free floating in the water, like plankton. See P. Araujo, M. Lemos, F. Mergulhão, L. Melo, and M. Simoes, "Antimicrobial Resistance to Disinfectants in Biofilms," in *Science against Microbial Pathogens: Communicating Current Research and Technological Advances*, ed. A. Mendez-Vilas, 826–834 (Badajoz: Formatex, 2011).

7. L. G. Wilson, "Commentary: Medicine, Population, and Tuberculosis," *International Journal of Epidemiology* 34, no. 3 (2004): 521–524.

8. K. I. Bos, K. M. Harkins, A. Herbig, M. Coscolla, N. Weber, I. Comas, S. A. Forrest, J. M. Bryant, S. R. Harris, V. J. Schuenemann, and T. J Campbell, "Pre-Columbian Mycobacterial Genomes Reveal Seals as a Source of New World Human Tuberculosis," *Nature* 514, no. 7523 (2014): 494–497. Also see S. Rodriguez-Campos, N. H. Smith, M. B. Boniotti, and A. Aranaz, "Overview and Phylogeny of *Mycobacterium tuberculosis* Complex Organisms: Implications for Diagnostics and Legislation of Bovine Tuberculosis," *Research in Veterinary Science* 97 (2014): S5–S19.

9. W. Hoefsloot, J. Van Ingen, C. Andrejak, K. Ängeby, R. Bauriaud, P. Bemer, N. Beylis, et al., "The Geographic Diversity of Nontuberculous Mycobacteria Isolated from Pulmonary Samples: An NTM-NET Collaborative Study," *European Respiratory Journal* 42, no. 6 (2013): 1604–1613.

10. J. R. Honda, N. A. Hasan, R. M. Davidson, M. D. Williams, L. E. Epperson, P. R. Reynolds, and E. D. Chan, "Environmental Nontuberculous Mycobacteria in the Hawaiian Islands," *PLoS Neglected Tropical Diseases* 10, no. 10 (2016): e0005068. See also an important early study on showerhead microbes: L. M. Feazel, L. K. Baumgartner, K. L. Peterson, D. N. Frank, J. K. Harris, and N. R. Pace, "Opportunistic Pathogens Enriched in Showerhead Biofilms," *Proceedings of the National Academy of Sciences* 106, no. 38 (2009): 16393–16399.

11. By that I mean that I sent Lauren Nichols in my lab an email and asked her to do it. Lauren sent the email to Lea Shell. Lea and Lauren would eventually share the email with Julie Sheard (a graduate student in our group based in Denmark).

12. The tenth time being when he tried to enlist me to sample my own urethral microbiome. No thanks.

13. In general, in your water, the more conducive conditions are to growth, the fewer species seem to be present. Flowing cold water is the most diverse, followed by flowing warm water, followed by stagnant water, followed by the biofilms, which are the least diverse. See figure 4b in C. R. Proctor, M. Reimann, B. Vriens, and F. Hammes, "Biofilms in Shower Hoses," *Water Research* 131 (2018): 274–286.

14. Modern universities are arranged into colleges (for example, my university has a College of Humanities and Social Sciences—CHASS—and a College of Agriculture and Life Sciences—CALS—as well as many others). Each college is run by a dean, just as each department is run by a head or a chair. But the dean does not act alone. She or he also has associate deans. The associate deans don't act alone. They have assistant deans. In some places, not even the assistant deans act alone. Just as each flea has lesser fleas, so too does each dean have lesser deans, aka deanlets.

15. E. Ludes and J. R. Anderson, "'Peat-Bathing' by Captive White-Faced Capuchin Monkeys (*Cebus capucinus*)," *Folia Primatologica* 65, no. 1 (1995): 38–42.

16. P. Zhang, K. Watanabe, and T. Eishi, "Habitual Hot Spring Bathing by a Group of Japanese Macaques (*Macaca fuscata*) in Their Natural Habitat," *American Journal of Primatology* 69, no. 12 (2007): 1425–1430.

17. Based on conversations with Hjalmar Kuehl at the Max Planck Institute in Leipzig. Kuehl and his colleagues have spent many hours watching chimpanzees.

18. Relative to hand washing and clean drinking water, bathing in a bath or in a shower has more to do with aesthetics and culture than it does with hygiene, to a point. When NASA was exploring the potential for extended space missions, they realized that the astronauts would need to spend long periods in the same piece of clothing. The astronauts were made to sit for days and then weeks, both in training and in real missions, without washing or changing their clothes. Their clothing deteriorated. Their skin developed boils. The sebum on their skin built into a dense layer and began to cake up. Which is to say, if you are washing your hands and keeping your bits clean, you don't have to shower or bathe very often, but you should do it more often than astronauts do, or at least more often than those astronauts did. See the chapter "Houston, We Have a Fungus" in M. Roach, *Packing for Mars: The Curious Science of Life in the Void* (New York: W. W. Norton, 2011).

19. See, for example, W. A. Fairservis, "The Harappan Civilization: New Evidence and More Theory," *American Museum Novitates*, no. 2055 (1961).

20. In what now seems like a very modern moment, the Roman emperor Commodus once staged a rigged battle between himself and an ostrich. The crowd was huge. The ostrich was tethered. Commodus was naked. Commodus proceeded to dispatch the ostrich and hold its head aloft to the senators sitting ringside, to great and thunderous applause, mostly. One of the senators, Dio, would go on to describe this moment as one of the most difficult in his life. It took incredible heroism for him to stifle his desire to giggle. He even took a laurel from the wreath he was wearing and put it in his mouth to keep from laughing aloud. See M. Beard, *Laughter in Ancient Rome: On Joking, Tickling, and Cracking Up* (Oakland: University of California Press, 2014).

21. G. G. Fagan, "Bathing for Health with Celsus and Pliny the Elder," *Classical Quarterly* 56, no. 1 (2006): 190–207.

22. An excavation of a latrine at a Roman bath in Sagalassos in what was then Asia Minor and is now Turkey revealed eggs of roundworms (*Ascaris* spp.) as well as evidence of the protist *Giardia duodenalis*. F. S. Williams, T. Arnold-Foster, H. Y. Yeh, M. L. Ledger, J. Baeten, J. Poblome, and P. D. Mitchell,

"Intestinal Parasites from the 2nd–5th Century AD Latrine in the Roman Baths at Sagalassos (Turkey)," *International Journal of Paleopathology* 19 (2017): 37–42.

23. In the early Renaissance, both in Italy and in northern Europe, paintings of nude men in water became popular. These scenes evoked earlier Roman and Greek depictions but were almost universally depictions of men swimming rather than men engaged in efforts to clean themselves. Among the exceptions is a print by Albrecht Dürer (1471–1528) in which Dürer depicts himself and three of his friends at a male bathhouse in Germany. Such bathhouses were used both for bathing oneself and for socializing, though perhaps as much the latter as the former as is suggested by the closing of bathhouses in Nuremberg just before Dürer's print was made because of the perceived spread of syphilis therein. See S. S. Dickey, "Rembrandt's 'Little Swimmers' in Context," in *Midwest Arcadia: Essays in Honor of Alison Kettering* (2015), doi:10.18277/makf.2015.05.

24. One exception was that of the Vikings. The Vikings were ferocious raiders of other peoples whose military successes depended on their ferocity, their weapons, and their very fast boats. They were also farmers. These two characteristics seem to be well known (and well documented). What is less well known is that the Vikings were also very fashion conscious. They used lye soap to bleach their hair before sailing out to conquer an abbey (much as modern Danes, descendants of those Vikings, bleach theirs before getting on their bikes to ride through Copenhagen). They also used lye soap on the rest of their bodies and on their clothes. As a result, it is likely that the Vikings' bodies and clothes had very different species than did those of their Dark Ages colleagues, including fewer body lice than, say, many an English queen.

25. F. Geels, "Co-evolution of Technology and Society: The Transition in Water Supply and Personal Hygiene in the Netherlands (1850–1930)—a Case Study in Multi-level Perspective," *Technology in Society* 27, no. 3 (2005): 363–397.

26. Yes, bottled water contains bacteria too. Learn to love them. S. C. Edberg, P. Gallo, and C. Kontnick, "Analysis of the Virulence Characteristics of Bacteria Isolated from Bottled, Water Cooler, and Tap Water," *Microbial Ecology in Health and Disease* 9, no. 2 (1996): 67–77. In some studies, bottled water has actually been found to contain a much higher density of bacteria than does tap water. J. A. Lalumandier and L. W. Ayers, "Fluoride and Bacterial Content of Bottled Water vs. Tap Water," *Archives of Family Medicine* 9, no. 3 (2000): 246.

27. Ninety-four percent of all the liquid (non-ice) freshwater on Earth is groundwater. C. Griebler and M. Avramov, "Groundwater Ecosystem Services: A Review," *Freshwater Science* 34, no. 1 (2014): 355–367.

28. The fates of viruses in a diverse aquifer are little better (some protists even break apart viruses and incorporate their amino acids into their cells).

29. For a great review of all the ways a pathogen can die in an aquifer, see J. Feichtmayer, L. Deng, and C. Griebler, "Antagonistic Microbial Interactions: Contributions and Potential Applications for Controlling Pathogens in the Aquatic Systems," *Frontiers in Microbiology* 8 (2017).

30. In a growing number of places (and ever more in our ever drier future), the treatment facility processes waste water and turns it, via a variety of ecological and chemical processes, into tap water.

31. F. Rosario-Ortiz, J. Rose, V. Speight, U. Von Gunten, and J. Schnoor, "How Do You Like Your Tap Water?" *Science* 351, no. 6276 (2016): 912–914.

32. We have talked in the lab about doing a water tasting to see which of these factors matters most to how people enjoy what they drink (and even which microbes might imbue water with special flavors). We haven't yet, but one could. Pause and savor the next water you drink. Think about whether it has a hint of being "aged in clay pipes" or even the "faint and fruity flavor of crustacean."

33. L. M. Feazel, L. K. Baumgartner, K. L. Peterson, D. N. Frank, J. L. Harris, and N. R. Pace, "Opportunistic Pathogens Enriched in Showerhead Biofilms," *Proceedings of the National Academy of Sciences* 106, no. 38 (2009): 16393–16399.

34. S. O. Reber, P. H. Siebler, N.C. Donner, J. T. Morton, D. G. Smith, J. M. Kopelman, K. R. Lowe, et al., "Immunization with a Heat-Killed Preparation of the Environmental Bacterium *Mycobacterium vaccae* Promotes Stress Resilience in Mice," *Proceedings of the National Academy of Sciences* 113, no. 22 (2016): E3130–E3139.

## CHAPTER 6

1. S. Nash, "The Plight of Systematists: Are They an Endangered Species?" October 16, 1989, https://www.the-scientist.com/?articles.view/articleNo/10690/title/The-Plight-Of-Systematists—Are-They-An-Endangered-Species-/. See also the more recent but similarly themed: L. W. Drew, "Are We Losing the Science of Taxonomy? As Need Grows, Numbers and Training Are Failing to Keep Up," *BioScience* 61, no. 12 (2011): 942–946.

2. The analysis of these data was a herculean task requiring patience, coding, vision, and then a little more patience. It was a task carried out by Albert Barberán, now at the University of Arizona, Tucson. See A. Barberán, R. R. Dunn, B. J. Reich, K. Pacifici, E. B. Laber, H. L. Menninger, J. M. Morton, et al., "The Ecology of Microscopic Life in Household Dust," *Proceedings of the Royal Society B: Biological Sciences* 282, no. 1814 (2015): 20151139. See also

A. Barberán, J. Ladau, J. W. Leff, K. S. Pollard, H. L. Menninger, R. R. Dunn, and N. Fierer, "Continental-Scale Distributions of Dust-Associated Bacteria and Fungi," *Proceedings of the National Academy of Sciences* 112, no. 18 (2015): 5756–5761. We would also ultimately be able to consider not just fungi but also mutualisms for which fungi are one of the key partners, including lichens. See E. A. Tripp, J. C. Lendemer, A. Barberán, R. R. Dunn, and N. Fierer, "Biodiversity Gradients in Obligate Symbiotic Organisms: Exploring the Diversity and Traits of Lichen Propagules across the United States," *Journal of Biogeography* 43, no. 8 (2016): 1667–1678.

3. Because no one on our team is trained as a fungal systematist, we actually can't name new species we find, even if we can grow them. Naming requires someone with Birgitte's skills, and people with Birgitte's skills tend to be very, very busy.

4. V. A. Robert and A. Casadevall, "Vertebrate Endothermy Restricts Most Fungi as Potential Pathogens," *Journal of Infectious Diseases* 200, no. 10 (2009): 1623–1626.

5. Many of the fungal species whose DNA we found in homes were probably dead. They drifted in. They landed. They then died, unable to cope with the hostile conditions in our bedrooms and kitchens. These fungi can't grow. They can't produce new compounds, metabolites, that make us sick. They can't produce more allergens. They are ghosts whose presence is detectable but of little consequence. Other species of fungi in homes are quiescent, hanging out as spores and waiting for the right conditions to grow, some perfect mélange of food and water or, in many cases, just the right amount of water.

6. N. S. Grantham, B. J. Reich, K. Pacifici, E. B. Laber, H. L. Menninger, J. B. Henley, A. Barberán, J. W. Leff, N. Fierer, and R. R. Dunn, "Fungi Identify the Geographic Origin of Dust Samples," *PLoS One* 10, no. 4 (2015): e0122605.

7. Though even this seemingly simple statement comes with a caveat. In one Russian study, species of household fungi and human skin bacteria exposed on the exterior of the International Space Station (yes, the exterior!) survived for at least thirteen months. See V. M. Baranov, N. D. Novikova, N. A. Polikarpov, V. N. Sychev, M. A. Levinskikh, V. R. Alekseev, T. Okuda, M. Sugimoto, O. A. Gusev, and A. I. Grigor'ev, "The Biorisk Experiment: 13-Month Exposure of Resting Forms of Organism on the Outer Side of the Russian Segment of the International Space Station: Preliminary Results," *Doklady Biological Sciences* 426, no. 1 (2009): 267–270. MAIK Nauka/Interperiodica.

8. For example, no cultures appear to have been done at the high temperatures necessary to grow thermophiles. Nor were any samples taken in such a way as to consider bacteria or fungi that are hard to culture for other reasons or even unculturable.

9. What was more, the fungi on Mir grew up to four times more rapidly than did their relatives on the ground. Why this might be remains a mystery. See N. D. Novikova, "Review of the Knowledge of Microbial Contamination of the Russian Manned Spacecraft," *Microbial Ecology* 47, no. 2 (2004): 127–132. The fungi also seemed to have cycles of some sort, though just why this might be (far beyond the reach of Earth's seasons) has not been explored. Novikova links these cycles to radiation levels experienced in the space station, but just why levels of radiation would affect the fungi in this way is unclear.

10. O. Makarov, "Combatting Fungi in Space," *Popular Mechanics*, January 1, 2016, 42–46.

11. Novikova, "Review of the Knowledge of Microbial Contamination of the Russian Manned Spacecraft," 127–132.

12. T. A. Alekhova, N. A. Zagustina, A. V. Aleksandrova, T. Y. Novozhilova, A. V. Borisov, and A. D. Plotnikov, "Monitoring of Initial Stages of the Biodamage of Construction Materials Used in Aerospace Equipment Using Electron Microscopy," *Journal of Surface Investigation: X-ray, Synchrotron and Neutron Techniques* 1, no. 4 (2007): 411–416.

13. Also found on Mir was *Botrytis*, a pathogen of grapes; it may have hopped a ride, alive, in wine.

14. It is distinct from the other pink bacteria in bathrooms, *Serratia marcescens*, which is more common in places that are always wet, such as toilet bowls. *Serratia* was also found on Mir. In both cases, the pink color of the bacteria is due to compounds that protect them from UV rays, a kind of bacterial sunscreen. *Rhodotorula* can also harvest nitrogen from the air, so is well suited to living in places that seem unlivable.

15 N. Novikova, P. De Boever, S. Poddubko, E. Deshevaya, N. Polikarpov, N. Rakova, I. Coninx, and M. Mergeay, "Survey of Environmental Biocontamination on Board the International Space Station," *Research in Microbiology* 157, no. 1 (2016): 5–12.

16. These included three *Candida* taxa, *Cryptococcus oeirensis*, *Penicillium concetricum*, and brewer's yeast (*Saccharomyces cerevisiae*). Also more common in houses with more people were *Rhodotorula mucilaginosa* and *Cystofilobasidium capitatum*, both of which are species that do well under stressful conditions such as those associated with bathrooms that are frequently cleaned.

17. Air conditioners were also associated with several other fungal species though, including the wood rot fungus *Physisporinus vitreus*, a pattern the mechanistic links of which deserve more study.

18. The more often you use your air-conditioning system, the more fungi build up in the AC unit. To avoid spreading these fungi through your house via the air conditioner, clean the filter with a vacuum or by hand washing with soap, which seems to help. Also, because AC units spread the most fungi

in the first ten minutes after they are turned on, some scientists recommend opening your windows each time you turn your AC on. Or you can keep the AC off and open the windows, which has the added advantage of allowing a rich biodiversity of environmental bacteria to blow in. N. Hamada and T. Fujita, "Effect of Air-Conditioner on Fungal Contamination," *Atmospheric Environment* 36, no. 35 (2002): 5443–5448.

19. I say "to my knowledge" because science projects often go up on the ISS and some of these might well contain cellulose and lignin. When he was a postdoc in my lab, Clint Penick worked with Eleanor Spicer Rice (my friend and neighbor) to collect pavement ants (*Tetramorium* sp.) that were later sent up to live on the ISS for a while. Those ants would have carried many North Carolinian fungi and bacteria with them, conceivably some able to break down cellulose and lignin.

20. In practice, there are multiple reasons it might have been rare in dust. It could be truly rare in homes. Or it could be rare for reasons having to do with the details of the science of sequencing. Neither of these phenomena would prove, however, to be the most interesting possibility.

21. She identified species of *Chaetomium*, *Penicillium*, *Mucor*, and *Aspergillus*.

22. Species of *Mucor* have been found not only in human homes but also in those of wasps, suggesting that the relationship of fungi to homes (including nests) may be far older than our species and date instead to the origin of wasp homes, tens of millions of years earlier. See A. A. Madden, A. M. Stchigel, J. Guarro, D. Sutton, and P. T. Starks, "*Mucor nidicola* sp. nov., a Fungal Species Isolated from an Invasive Paper Wasp Nest," *International Journal of Systematic and Evolutionary Microbiology* 62, no. 7 (2012): 1710–1714. For a beautiful study of the evolution of the architecture of wasp nests, see R. L. Jeanne, "The Adaptiveness of Social Wasp Nest Architecture," *Quarterly Review of Biology* 50, no. 3 (1975): 267–287.

23. *Chaetomium* was found growing on surfaces in Mir, but not in the air. *Penicillium* species were everywhere in Mir (in nearly 80 percent of samples). *Mucor* was in 1–2 percent of Mir samples. *Aspergillus* was in 40 percent of surface samples on Mir and 76.6 percent of air samples.

24. P. F. E. W. Hirsch, F. E. W. Eckhardt, and R. J. Palmer Jr., "Fungi Active in Weathering of Rock and Stone Monuments," *Canadian Journal of Botany* 73, no. S1 (1995): 1384–1390.

25. Most termites cannot break down lignin but get around this problem by dragging with them, wherever they go, guts full of bacteria and protists that can do the job. In nature, the work of termites and their microbes is integral to the existence of forests and grasslands. Termites speed up decomposition in ways that make trees grow faster and grasses grow taller and more generally maintain healthy, functioning ecosystems. But when we build homes, we want

to forestall these processes (and termites) for as long as we can, much as we try to do the same in keeping fruit or meat sound until we eat it.

26. Among them were *Arthrinium phaeospermum*, *Aureobasidium pullulans*, *Cladosporium herbarum*, species of *Trichoderma*, *Alternaria tenuissima*, species of *Fusarium*, species of *Gliocladium*, *Rhodotorula mucilaginosa*, and *Trichosporon pullulans*. Few of these fungi were present on the space station or on Mir, which is perhaps unsurprising given that not much on the space station is made of wood.

27. H. Kauserud, H. Knudsen, N. Högberg, and I. Skrede, "Evolutionary Origin, Worldwide Dispersal, and Population Genetics of the Dry Rot Fungus *Serpula lacrymans*," *Fungal Biology Reviews* 26, nos. 2–3 (2012): 84–93.

28. Among them *Penicillium*, *Chaetomium*, and *Ulocladium*.

29. R. I. Adams, M. Miletto, J. W. Taylor, and T. D. Bruns, "Dispersal in Microbes: Fungi in Indoor Air Are Dominated by Outdoor Air and Show Dispersal Limitation at Short Distances," *ISME Journal* 7, no. 7 (2013): 1262–1273.

30. D. L. Price and D. G. Ahearn, "Sanitation of Wallboard Colonized with *Stachybotrys chartarum*," *Current Microbiology* 39, no. 1 (1999): 21–26.

31. They have also seen stories like that of Tyrone Hayes. Tyrone studies the effects of an herbicide on animals. He found that the herbicide harms animals. The result, as Rachel Aviv describes in the *New Yorker*, is that "its maker pursued him," and not in a good way (see "A Valuable Reputation," February 10, 2014, www.newyorker.com/magazine/2014/02/10/a-valuable-reputation).

32. Birgitte is fascinated by *Chaetomium* species. They have, as she told me in an email, always surrounded her. For example, she sent me a photo of her elementary school class from when she was a little girl. The photo was labeled with an arrow. The arrow pointed not to Birgitte but instead to the fungus *Chaetomium elatum*, growing on the paper on which the photo was mounted.

33. Interestingly, none of these species was found on the International Space Station or even on the much more fungal Mir.

34. M. Nikulin, K. Reijula, B. B. Jarvis, and E.-L. Hintikka, "Experimental Lung Mycotoxicosis in Mice Induced by *Stachybotrys atra*," *International Journal of Experimental Pathology* 77, no. 5 (1996): 213–218.

35. I. Došen, B. Andersen, C. B. W. Phippen, G. Clausen, and K. F. Nielsen, "*Stachybotrys* Mycotoxins: From Culture Extracts to Dust Samples," *Analytical and Bioanalytical Chemistry* 408, no. 20 (2016): 5513–5526.

36. *Alternaria alternate*, *Aspergillus fumigatus*, and *Cladosporium herbarum* are all among the species present in Birgitte's study, present, too, on the space station. These fungi are all commonly associated with allergies.

37. A. Nevalainen, M. Täubel, and A. Hyvärinen, "Indoor Fungi: Companions and Contaminants," *Indoor Air* 25, no. 2 (2015): 125–156.

38. C. M. Kercsmar, D. G. Dearborn, M. Schluchter, L. Xue, H. L. Kirchner, J. Sobolewski, S. J. Greenberg, S. J. Vesper, and T. Allan, "Reduction in Asthma Morbidity in Children as a Result of Home Remediation Aimed at Moisture Sources," *Environmental Health Perspectives* 114, no. 10 (2006): 1574.

## CHAPTER 7

1. Though probably mostly in defense. Most recent studies argue that cave bears were largely herbivorous. But, from the perspective of a small human, a big, pissed-off herbivorous bear trapped in a cave is, first and foremost, just a big, pissed-off bear.

2. It would be a great story if the camel crickets were named for Francois Camel, who led the boys to the cave. I'm happy to pretend this is true if you are. The truth, though, is that camel crickets are called camel crickets because their backs are arched, like a camel's hump.

3. Camel crickets are no longer found in the French Pyrenees, which raises another question: Where did the early human who depicted this camel cricket see the camel cricket? One option is that the camel crickets used to live in the French Pyrenees, but no longer do. This is possible, but unlikely. Caves in France at the time would have been much colder than they are today, and the modern distribution of *Troglophilus* camel crickets does not include France. The crickets occur only much farther south. Another possibility is that the artist saw the camel cricket in a more southerly cave and was depicting it from memory. Or maybe the artist actually made the art elsewhere and carried it with him or her.

4. S. Hubbell, *Broadsides from the Other Orders* (New York: Random House, 1994).

5. The cascade of resources from cricket food to cricket predators can be complex and bizarre. See, for example, the special case of hairworms, which can take control of the body and will of a camel cricket. T. Sato, M. Arizono, R. Sone, and Y. Harada, "Parasite-Mediated Allochthonous Input: Do Hairworms Enhance Subsidized Predation of Stream Salmonids on Crickets?" *Canadian Journal of Zoology* 86, no. 3 (2008): 231–235. See also: Y. Saito, I. Inoue, F. Hayashi, and H. Itagaki, "A Hairworm, *Gordius* sp., Vomited by a Domestic Cat," *Nihon Juigaku Zasshi: The Japanese Journal of Veterinary Science* 49, no. 6 (1987): 1035–1037.

6. She is also, as one of her reference letters pointed out, a very talented fiddle player. You can hear her playing here: https://youtu.be/aVXG5koU9G4.

7. The parade-like scene of the team entering houses was not without precedent. Linnaeus, the father of modern taxonomy who named many of the more common arthropod species found in houses, actually did have

a band that paraded before him when he went on excursions. The drum that was played as they walked together has been preserved. See B. Jonsell, "Daniel Solander—the Perfect Linnaean; His Years in Sweden and Relations with Linnaeus," *Archives of Natural History* 11, no. 3 (1984): 443–450.

8. Entomologists spend a lot of time looking at the genitalia of insects. This reality, combined with the unique ways in which entomologists show their love and appreciation, can lead to unusual circumstances. For example, my friend Dan Simberloff recently had a new species of louse from a swiftlet named in his honor. One cannot but be flattered by something like this, and yet one should also note that the characteristics that make this new louse species, *Dennyus simberloffi*, unique and distinguishable from its closest relatives are its unusually small genitalia and very wide head and anus. See D. Clayton, R. Price, and R. Page, "Revision of *Dennyus* (*Collodennyus*) Lice (Phthiraptera: Menoponidae) from Swiftlets, with Descriptions of New Taxa and a Comparison of Host–Parasite Relationships," *Systematic Entomology* 21, no. 3 (1996): 179–204.

9. If there is an afterlife for entomologists, it involves being kept in a jar for a while until some overworked God decides whether or not they are in good enough condition to be pinned.

10. A. A. Madden, A. Barberán, M. A. Bertone, H. L. Menninger, R. R. Dunn, and N. Fierer, "The Diversity of Arthropods in Homes across the United States as Determined by Environmental DNA Analyses," *Molecular Ecology* 25, no. 24 (2016): 6214–6224.

11. This relationship between wasp and aphid was first observed by Leeuwenhoek on an aphid just outside his house in Delft. See F. N. Egerton, "A History of the Ecological Sciences, Part 19: Leeuwenhoek's Microscopic Natural History," *Bulletin of the Ecological Society of America* 87 (2006): 47–58.

12. See, for example, E. Panagiotakopulu, "New Records for Ancient Pests: Archaeoentomology in Egypt," *Journal of Archaeological Science* 28, no. 11 (2001): 1235–1246; E. Panagiotakopulu, "Hitchhiking across the North Atlantic—Insect Immigrants, Origins, Introductions and Extinctions," *Quaternary International* 341 (2014): 59–68; E. Panagiotakopulu, P. C. Buckland, and B. J. Kemp, "Underneath Ranefer's Floors—Urban Environments on the Desert Edge," *Journal of Archaeological Science* 37, no. 3 (2010): 474–481; E. Panagiotakopulu and P. C. Buckland, "Early Invaders: Farmers, the Granary Weevil and Other Uninvited Guests in the Neolithic," *Biological Invasions* 20, no. 1 (2018): 219–233.

13. A. Bain, "A Seventeenth-Century Beetle Fauna from Colonial Boston," *Historical Archaeology* 32, no. 3 (1998): 38–48.

14. E. Panagiotakopulu, "Pharaonic Egypt and the Origins of Plague," *Journal of Biogeography* 31, no. 2 (2004): 269–275.

15. For more on this story, see J. B. Johnson and K. S. Hagen, "A Neuropterous Larva Uses an Allomone to Attack Termites," *Nature* 289 (5797): 506.

16. E. A. Hartop, B. V. Brown, R. Henry, and L. Disney, "Opportunity in Our Ignorance: Urban Biodiversity Study Reveals 30 New Species and One New Nearctic Record for Megaselia (Diptera: Phoridae) in Los Angeles (California, USA)," *Zootaxa* 3941, no. 4 (2015): 451–484.

17. E. A. Hartop, B. V. Brown, R. Henry, and L. Disney, "Flies from LA, the Sequel: A Further Twelve New Species of Megaselia (Diptera: Phoridae) from the BioSCAN Project in Los Angeles (California, USA)," *Biodiversity Data Journal* 4 (2016).

18. J. A. Feinberg, C. E. Newman, G. J. Watkins-Colwell, M. D. Schlesinger, B. Zarate, B. R. Curry, H. B. Shaffer, and J. Burger, "Cryptic Diversity in Metropolis: Confirmation of a New Leopard Frog Species (Anura: Ranidae) from New York City and Surrounding Atlantic Coast Regions," *PLoS One* 9, no. 10 (2014): e108213; J. Gibbs, "Revision of the Metallic *Lasioglossum* (Dialictus) of Eastern North America (Hymenoptera: Halictidae: Halictini)," *Zootaxa* 3073 (2011): 1–216; D. Foddai, L. Bonato, L. A. Pereira, and A. Minelli, "Phylogeny and Systematics of the Arrupinae (Chilopoda Geophilomorpha Mecistocephalidae) with the Description of a New Dwarfed Species," *Journal of Natural History* 37 (2003): 1247–1267, https://doi.org/10.1080/00222930210121672.

19. Y. Ang, G. Rajaratnam, K. F. Y. Su, and R. Meier, "Hidden in the Urban Parks of New York City: *Themira lohmanus*, a New Species of Sepsidae Described Based on Morphology, DNA Sequences, Mating Behavior, and Reproductive Isolation (Sepsidae, Diptera)," *ZooKeys* 698 (2017): 95.

20. In the book H. W. Greene, *Tracks and Shadows: Field Biology as Art* (Berkeley: University of California Press, 2013).

21. See I. Kant, *Critique of Judgment. 1790*, trans. W. S. Pluhar (Indianapolis: Hackett 212, 1987).

## CHAPTER 8

1. Another characteristic of cave organisms is their ability to go long periods without food. One ethnographer found silverfish (a species of *Lepisma*, which we also found to be common in Raleigh) to be abundant in Zulu houses. Out of interest, he trapped one of these silverfish in a wine glass, and it lived for at least three months with nothing for nutrition but the dust beneath the glass. See L. Grout, *Zulu-Land; or, Life among the Zulu-Kafirs of Natal and Zulu-Land, South Africa* (London: Trübner & Co., 1860).

2. See A. J. De Jesús, A. R. Olsen, J. R. Bryce, and R. C. Whiting, "Quantitative Contamination and Transfer of *Escherichia coli* from Foods by Houseflies, *Musca domestica* L. (Diptera: Muscidae)," *International Journal of Food Microbiology* 93, no. 2 (2004): 259–262. See also N. Rahuma, K. S. Ghenghesh, R. Ben Aissa, and A. Elamaari, "Carriage by the Housefly (*Musca domestica*) of Multiple-Antibiotic-Resistant Bacteria That Are Potentially Pathogenic to Humans, in Hospital and Other Urban Environments in Misurata, Libya," *Annals of Tropical Medicine and Parasitology* 99, no. 8 (2005): 795–802.

3. Evolutionary biologists call these "primary endosymbioses" to distinguish them from secondary endosymbioses in which the bacteria (the symbionts) are picked up later.

4. J. J. Wernegreen, S. N. Kauppinen, S. G. Brady, and P. S. Ward, "One Nutritional Symbiosis Begat Another: Phylogenetic Evidence That the Ant Tribe Camponotini Acquired *Blochmannia* by Tending Sap-Feeding Insects," *BMC Evolutionary Biology* 9, no. 1 (2009): 292; R. Pais, C. Lohs, Y. Wu, J. Wang, and S. Aksoy, "The Obligate Mutualist *Wigglesworthia glossinidia* Influences Reproduction, Digestion, and Immunity Processes of Its Host, the Tsetse Fly," *Applied and Environmental Microbiology* 74, no. 19 (2008): 5965–5974. Also see G. A. Carvalho, A. S. Corrêa, L. O. de Oliveira, and R. N. C. Guedes, "Evidence of Horizontal Transmission of Primary and Secondary Endosymbionts between Maize and Rice Weevils (*Sitophilus zeamais* and *Sitophilus oryzae*) and the Parasitoid *Theocolax elegans*," *Journal of Stored Products Research* 59 (2014): 61–65. Also see A. Heddi, H. Charles, C. Khatchadourian, G. Bonnot, and P. Nardon, "Molecular Characterization of the Principal Symbiotic Bacteria of the Weevil *Sitophilus oryzae*: A Peculiar G+ C Content of an Endocytobiotic DNA," *Journal of Molecular Evolution* 47, no. 1 (1998): 52–61.

5. C. M. Theriot and A. M. Grunden, "Hydrolysis of Organophosphorus Compounds by Microbial Enzymes," *Applied Microbiology and Biotechnology* 89, no. 1 (2011): 35–43.

6. The species was *Paenibacillus glucanolyticus* SLM1. Stephanie and Amy isolated this species from old and abandoned black liquor storage tanks kept in the demonstration paper-pulping facility at North Carolina State University. Yes, the university has a demonstration paper-pulping facility.

7. And a strong belief in the ability of nature, particularly bacterial nature, to solve problems.

8. And one could also check the many non-arthropod invertebrates, such as microscopic nematodes. It is said that the microscopic worms, the nematodes, in homes are so dense that if you removed the structure of a home and could make the worms visible, you would see the home still, framed in the outline of serpentine bodies. This could be the case. Yet, we were able to find no studies of nematodes in homes, nor the tardigrades, nor many other major

groups of organisms. These creatures are there, but have not yet been tallied, much less had their potential uses considered.

9. F. Sabbadin, G. R. Hemsworth, L. Ciano, B. Henrissat, P. Dupree, T. Tryfona, R. D. S. Marques, et al., "An Ancient Family of Lytic Polysaccharide Monooxygenases with Roles in Arthropod Development and Biomass Digestion," *Nature Communications* 9, no. 1 (2018): 756.

10. T. D. Morgan, P. Baker, K. J. Kramer, H. H. Basibuyuk, and D. L. J. Quicke, "Metals in Mandibles of Stored Product Insects: Do Zinc and Manganese Enhance the Ability of Larvae to Infest Seeds?" *Journal of Stored Products Research* 39, no. 1 (2003): 65–75.

11. Coby Schal and Ayako Wada-Katsumata at North Carolina State University have also teamed up to study the subset of these brushes insects use to clean their antennae. They discovered that when insects such as carpenter ants (*Camponotus pennsylvanicus*), house flies, and German cockroaches clean their antennae, they are better able to smell. With dirty antennae, the world is dulled. See K. Böröczky, A. Wada-Katsumata, D. Batchelor, M. Zhukovskaya, and C. Schal, "Insects Groom Their Antennae to Enhance Olfactory Acuity," *Proceedings of the National Academy of Sciences* 110, no. 9 (2013): 3615–3620.

12. E. L. Zvereva, "Peculiarities of Competitive Interaction between Larvae of the House Fly *Musca domestica* and Microscopic Fungi," *Zoologicheskii Zhurnal* 65 (1986): 1517–1525. See also K. Lam, K. Thu, M. Tsang, M. Moore, and G. Gries, "Bacteria on Housefly Eggs, *Musca domestica*, Suppress Fungal Growth in Chicken Manure through Nutrient Depletion or Antifungal Metabolites," *Naturwissenschaften* 96 (2009): 1127–1132.

13. D. A. Veal, Jane E. Trimble, and A. J. Beattie, "Antimicrobial Properties of Secretions from the Metapleural Glands of *Myrmecia gulosa* (the Australian Bull Ant)," *Journal of Applied Microbiology* 72, no. 3 (1992): 188–194.

14. C. A. Penick, O. Halawani, B. Pearson, S. Mathews, M. M. López-Uribe, R. R. Dunn, and A. A. Smith, "External Immunity in Ant Societies: Sociality and Colony Size Do Not Predict Investment in Antimicrobials," *Royal Society Open Science* 5, no. 2 (2018): 171332.

15. I. Stefanini, L. Dapporto, J.-L. Legras, A. Calabretta, M. Di Paola, C. De Filippo, R. Viola, et al. "Role of Social Wasps in *Saccharomyces cerevisiae* Ecology and Evolution," *Proceedings of the National Academy of Sciences* 109, no. 33 (2012): 13398–13403.

16. This work was made possible thanks to Anne Madden's ability to know, find, listen to, and sniff out new and interesting yeasts and John Sheppard's ability to brew beer. For more on this project, see www.pbs.org/newshour/bb/wing-wasp-scientists-discover-new-beer-making-yeast/.

17. A. Madden, MJ Epps, T. Fukami, R. E. Irwin, J. Sheppard, D. M. Sorger, and R. R. Dunn, "The Ecology of Insect–Yeast Relationships and Its

Relevance to Human Industry," *Proceedings of the Royal Society B* 285, no. 1875 (2018): 20172733.

18. E. Panagiotakopulu, "Dipterous Remains and Archaeological Interpretation," *Journal of Archaeological Science* 31, no. 12 (2004): 1675–1684.

19. E. Panagiotakopulu, P. C. Buckland, P. M. Day, and C. Doumas, "Natural Insecticides and Insect Repellents in Antiquity: A Review of the Evidence," *Journal of Archaeological Science* 22, no. 5 (1995): 705–710.

## CHAPTER 9

1. R. E. Heal, R. E. Nash, and M. Williams, "An Insecticide-Resistant Strain of the German Cockroach from Corpus Christi, Texas," *Journal of Economic Entomology* 46, no. 2 (1953).

2. As with fipronil, the active compound in some cockroach baits and also in some flea sprays/powders/pills. See G. L. Holbrook, J. Roebuck, C. B. Moore, M. G. Waldvogel, and C. Schal, "Origin and Extent of Resistance to Fipronil in the German Cockroach, *Blattella germanica* (L.) (Dictyoptera: Blattellidae)," *Journal of Economic Entomology* 96, no. 5 (2003): 1548–1558.

3. These pesticides were so potent that they posed risks to birds and children alike (particularly in the concentrations at which they were being used); they were the pesticides about which Rachel Carson wrote in her documentary *Silent Spring*. Yet they were not potent enough to kill all of the German cockroaches.

4. Yes, Pleasanton is the name of the place where one goes to study cockroaches and other pest species. In Pleasanton, Jules had already spent three years studying another pest, the cat flea (*Ctenocephalides felis*). Cat fleas were already present in human homes in ancient Amarna, Egypt. Jules discovered that cat flea larvae subsist on the bloody feces of their parents, supplemented, it seems, by microbes from the environment, microbes that add nutrition to the feces. See J. Silverman and A. G. Appel, "Adult Cat Flea (Siphonaptera: Pulicidae) Excretion of Host Blood Proteins in Relation to Larval Nutrition," *Journal of Medical Entomology* 31, no. 2 (1993): 265–271.

5. Most of these common names for the cockroaches now bear little relation to what we have learned about their histories. The American cockroach, for instance, appears to be native to Africa. The Oriental cockroach also appears to have been African and may have traveled with the Phoenicians, then the Greeks, then nearly everyone else. See R. Schweid, *The Cockroach Papers: A Compendium of History and Lore* (Chicago: University of Chicago Press, 2015). For a classic, see J. A. G. Rehn, "Man's Uninvited Fellow Traveler—the Cockroach," *Scientific Monthly* 61 no. 145 (1945): 265–276.

6. These species are incredibly varied in their lifeways. Many species of wild cockroaches are diurnal, active during the day. They often feed on leaf litter in the forest. Quite a few live in the nests of ants and termites as guests. Some even produce a kind of mother's milk on which their babies feed. Others pollinate flowers. What is more, recent studies have confirmed that termites are actually all a special branch of the cockroach evolutionary tree in which sociality evolved. Termites are social cockroaches. See R. R. Dunn, "Respect the Cockroach," *BBC Wildlife* 27, no. 4 (2009): 60.

7. From the Greek *Parthenos* for "virgin" and *genesis* for "creation."

8. The Surinam cockroach (*Pycnoscelus surinamensis*) has taken this to an extreme. No males of this species have ever been found in the wild. In lab colonies, males are sometimes born but are so dysfunctional they quickly die.

9. Of course there are some bad things German cockroaches do that we don't. German cockroaches have been reported to eat nearly anything and everything that contains starch, including cereals, stamps, drapes, book bindings, and paste.

10. German cockroaches, unlike many other cockroach species, do not do well when alone. They suffer from what is called isolation syndrome, which to me sounds a lot like some mix of loneliness and modest existential despair. When left alone, their metamorphosis is delayed, as is their sexual maturation. Also, they start to behave unusually, as if they don't know quite how to be a cockroach anymore. They are no longer interested in normal cockroach activities or even cockroach sex. There is a big literature on the loneliness of German cockroaches, but for a start, read M. Lihoreau, L. Brepson, and C. Rivault, "The Weight of the Clan: Even in Insects, Social Isolation Can Induce a Behavioural Syndrome," *Behavioural Processes* 82, no. 1 (2009): 81–84.

11. Half of the fifty or so species of *Blattella* live in Asia.

12. It may have happened with the earliest agriculture in tropical Asia. But it may also have been much later.

13. The oldest specimen of the German cockroach is actually from Denmark, so we can blame the Danes, but my suspicion is that the German cockroach actually arrived in Europe much, much earlier. See T. Qian, "Origin and Spread of the German Cockroach, *Blattella germanica*" (PhD diss., National University of Singapore, 2016).

14. Though the cockroach gets a little comeuppance. When the name of the German cockroach is written out in full, it is actually *Blattella germanica* Linnaeus. *Linnaeus* is placed after the species name to indicate that Linnaeus was the one who bestowed the name. This convention, along with the convention of giving each species a genus name (*Blattella*) and a species name (*germanica*), was invented by Linnaeus. As a result, wherever the German cockroach goes, forever, Linnaeus will drag behind it, as is the case with bed bugs, house flies,

black rats (*Cimex lectularis* Linnaeus, *Musca domestica* Linnaeus, and *Rattus rattus* Linnaeus, respectively), and many other indoor species.

15. P. J. A. Pugh, "Non-indigenous Acari of Antarctica and the Sub-Antarctic Islands," *Zoological Journal of the Linnaean Society* 110, no. 3 (1994): 207–217.

16. Which of the other species of cockroach are found indoors depends greatly on the climate outside and on geography. Some cockroach species do better in tropical environments, others where it is cold.

17. L. Roth and E. Willis, *The Biotic Association of Cockroaches*, Smithsonian Miscellaneous Collections, vol. 141 (Washington, DC: Smithsonian Institution, 1960).

18. Qian, "Origin and Spread of the German Cockroach."

19. J. Silverman and D. N. Bieman, "Glucose Aversion in the German Cockroach, *Blattella germanica*," *Journal of Insect Physiology* 39, no. 11 (1993): 925–933.

20. These rates of reproduction tend to outpace the rate at which new cockroaches move among buildings, so much so that German cockroaches of one lineage might occupy one apartment building and those of another might occupy the next.

21. J. Silverman and R. H. Ross, "Behavioral Resistance of Field-Collected German Cockroaches (Blattodea: Blattellidae) to Baits Containing Glucose," *Environmental Entomology* 23, no. 2 (1994): 425–430.

22. For example, see J. Silverman and D. N. Bieman, "High Fructose Insecticide Bait Compositions," US Patent No. 5,547,955 (1996).

23. See S. B. Menke, W. Booth, R. R. Dunn, C. Schal, E. L. Vargo, and J. Silverman, "Is It Easy to Be Urban? Convergent Success in Urban Habitats among Lineages of a Widespread Native Ant," *PLoS One* 5, no. 2 (2010): e9194.

24. See S. Lengyel, A. D. Gove, A. M. Latimer, J. D. Majer, and R. R. Dunn, "Ants Sow the Seeds of Global Diversification in Flowering Plants," *PLoS One* 4, no. 5 (2009): e5480. Also see S. Lengyel, A. D. Gove, A. M. Latimer, J. D. Majer, and R. R. Dunn, "Convergent Evolution of Seed Dispersal by Ants, and Phylogeny and Biogeography in Flowering Plants: A Global Survey," *Perspectives in Plant Ecology, Evolution and Systematics* 12, no. 1 (2010): 43–55. Also, for a quirky elaboration on this theme involving stick insects, see L. Hughes and M. Westoby, "Capitula on Stick Insect Eggs and Elaiosomes on Seeds: Convergent Adaptations for Burial by Ants," *Functional Ecology* 6, no. 6 (1992): 642–648.

25. In *Faust: A Tragedy*, Johann Wolfgang von Goethe has a demon introduce himself as "The lord of rats and the eke of mice, Of flies and bed-bugs, frogs and lice." Except for the frogs, this seems an apt title for the actions of

natural selection in our modern homes. See J. W. Goethe, *Faust: A Tragedy*, trans. B. Taylor (Boston: Houghton Mifflin, 1898), 1:86.

26. V. Markó, B. Keresztes, M. T. Fountain, and J. V. Cross, "Prey Availability, Pesticides and the Abundance of Orchard Spider Communities," *Biological Control* 48, no. 2 (2009): 115–124. Also see L. W. Pisa, V. Amaral-Rogers, L. P. Belzunces, J. M. Bonmatin, C. A. Downs, D. Goulson, D. P. Kreutzweiser, et al., "Effects of Neonicotinoids and Fipronil on Non-target Invertebrates," *Environmental Science and Pollution Research* 22, no. 1 (2015): 68–102.

27. We aren't the first species to imagine using predators in our homes to control pests. Many species that build nests benefit from other species that live in those nests. Some owls bring snakes to their nests to control the insects that eat their nestlings. Similarly, packrat nests often include pseudoscorpions that feed upon the mites that plague the packrats. See F. R. Gehlbach and R. S. Baldridge, "Live Blind Snakes (*Leptotyphlops dulcis*) in Eastern Screech Owl (*Otus asio*) Nests: A Novel Commensalism," *Oecologia* 71, no. 4 (1987): 560–563. Also see O. F. Francke and G. A. Villegas-Guzmán, "Symbiotic Relationships between Pseudoscorpions (Arachnida) and Packrats (Rodentia)," *Journal of Arachnology* 34, no. 2 (2006): 289–298.

28. O. F. Raum, *The Social Functions of Avoidances and Taboos among the Zulu*, vol. 6 (Berlin: Walter de Gruyter, 1973). The practice was then copied by the Voortrekkers, Boer pastoralists who arrived in the Cape Town region of South Africa with the Dutch East India Company and then migrated north and east in "great treks" in response to grievances with the British colonial government.

29. J. J. Steyn, "Use of Social Spiders against Gastro-intestinal Infections Spread by House Flies," *South African Medical Journal* 33 (1959).

30. J. Wesley Burgess, "Social spiders." *Scientific American* 234, no. 3 (1976): 100–107. This very cool spider appears to use dead flies in its web as a surface and food to farm yeasts, which then, in turn, attract live flies. No one has yet identified or even studied the yeast. W. J. Tietjen, L. R. Ayyagari, and G. W. Uetz, "Symbiosis between Social Spiders and Yeast: The Role in Prey Attraction," *Psyche* 94, nos. 1–2 (1987): 151–158.

31. Social spiders are confined in their distribution (French introduction attempts notwithstanding) and are not for everyone. Not to worry, there are other options. Jumping spiders in houses in Thailand eat up to 120 *Aedes* mosquitoes, vectors of deadly dengue fever, a day. See R. Weterings, C. Umponstira, and H. L. Buckley, "Predation on Mosquitoes by Common Southeast Asian House-Dwelling Jumping Spiders (Salticidae)," *Arachnology* 16, no. 4 (2014): 122–127. In Kenya, another spider that lives in houses preferentially feeds on the *Anopheles* mosquitoes that transmit malaria, particularly those that have already fed (and hence stand a higher chance of transmitting

malaria). See R. R. Jackson and F. R. Cross, "Mosquito-Terminator Spiders and the Meaning of Predatory Specialization," *Journal of Arachnology* 43, no. 2 (2015): 123–142. Also, see X. J. Nelson, R. R. Jackson, and G. Sune, "Use of Anopheles-Specific Prey-Capture Behavior by the Small Juveniles of *Evarcha culicivora*, a Mosquito-Eating Jumping Spider," *Journal of Arachnology* 33, no. 2 (2005): 541–548. Also, see X. J. Nelson and R. R. Jackson, "A Predator from East Africa That Chooses Malaria Vectors as Preferred Prey," *PLoS One* 1, no. 1 (2006): e132.

32. G. L. Piper, G. W. Frankie, and J. Loehr, "Incidence of Cockroach Egg Parasites in Urban Environments in Texas and Louisiana," *Environmental Entomology* 7, no. 2 (1978): 289–293.

33. A. M. Barbarin, N. E. Jenkins, E. G. Rajotte, and M. B. Thomas, "A Preliminary Evaluation of the Potential of *Beauveria bassiana* for Bed Bug Control," *Journal of Invertebrate Pathology* 111, no. 1 (2012): 82–85. And other labs are trying other fungi, be it on the common bed bug or its tropical relative, *Cimex hemipterus*. For example, see Z. Zahran, N. M. I. M. Nor, H. Dieng, T. Satho, and A. H. A. Majid, "Laboratory Efficacy of Mycoparasitic Fungi (*Aspergillus tubingensis* and *Trichoderma harzianum*) against Tropical Bed Bugs (*Cimex hemipterus*) (Hemiptera: Cimicidae)," *Asian Pacific Journal of Tropical Biomedicine* 7, no. 4 (2017): 288–293. Meanwhile, in Denmark, a parasitoid that attacks the pupae of house flies is being bred and released experimentally in stables in which dairy cattle are kept. This is being done in an attempt to control house flies as well as stable flies and to prevent their spillover into nearby homes. See H. Skovgård and G. Nachman, "Biological Control of House Flies *Musca domestica* and Stable Flies *Stomoxys calcitrans* (Diptera: Muscidae) by Means of Inundative Releases of *Spalangia cameroni* (Hymenoptera: Pteromalidae)," *Bulletin of Entomological Research* 94, no. 6 (2004): 555–567.

34. D. R. Nelsen, W. Kelln, and W. K. Hayes, "Poke but Don't Pinch: Risk Assessment and Venom Metering in the Western Black Widow Spider, *Latrodectus Hesperus*," *Animal Behaviour* 89 (2014): 107–114.

35. As for just how unlikely spiders are to bite, a recent case is illustrative. In Lenexa, Kansas, 2,055 brown recluse spiders (*Loxosceles reclusa*) were removed from an old house over a six-month period. No bites occurred in this or other houses with large populations of brown recluses. Thousands of spiders, but no bites. Meanwhile, most of the brown recluse bites reported in the United States occur in regions where the spider does not occur (which is to say these were not brown recluse bites and were very unlikely to even be spider bites). See R. S. Vetter and D. K. Barger, "An Infestation of 2,055 Brown Recluse Spiders (Araneae: Sicariidae) and No Envenomations in a Kansas Home: Implications for Bite Diagnoses in Nonendemic Areas," *Journal of Medical Entomology* 39, no. 6 (2002): 948–951.

36. M. H. Lizée, B. Barascud, J.-P. Cornec, and L. Sreng, "Courtship and Mating Behavior of the Cockroach *Oxyhaloa deusta* [Thunberg, 1784] (Blaberidae, Oxyhaloinae): Attraction Bioassays and Morphology of the Pheromone Sources," *Journal of Insect Behavior* 30, no. 5 (2017): 1–21.

37. Coby has identified this odor. He hasn't yet figured out, though, how to make large quantities of it. When he does, stay clear of Coby, because if he spills any of the stuff on him, he will be like the Pied Piper of German cockroaches.

38. A. Wada-Katsumata, J. Silverman, and C. Schal, "Changes in Taste Neurons Support the Emergence of an Adaptive Behavior in Cockroaches," *Science* 340 (2013): 972–975.

39. None of the disasters humanity has imagined—not nuclear war, not even the most extreme climate change—will end life. As Sean Nee has noted, all of the horrible things we have done to the planet, changes that disfavor many species, including those on which we depend, nonetheless favor a subset of unusual microbes. What deforestation, climate change, nuclear disaster, and the like tip us toward is a world in which the microbes, once again, more fully assert themselves, a world like that in the very beginning, a world of bountiful slime. See S. Nee, "Extinction, Slime, and Bottoms," *PLoS Biology* 2, no. 8 (2004): e272.

## CHAPTER 10

1. Here is part of Jim's thesis, if you are curious: J. A. Danoff-Burg, "Evolving under Myrmecophily: A Cladistic Revision of the Symphilic Beetle Tribe Sceptobiini (Coleoptera: Staphylinidae: Aleocharinae)," *Systematic Entomology* 19, no. 1 (1994): 25–45.

2. The units evolutionary biologists use to consider the benefits of one species to another species and, in doing so, to decide whether a relationship is parasitic or mutualistic are always units of Darwinian fitness. One species benefits another if it makes the second species more likely to survive and have more offspring that survive. Maybe this amoral economics of natural selection is no longer what we should use to consider which species benefit us and which fail to do so. Perhaps a species that makes us happy and "well," whatever "well" means, but that does not benefit our fitness is, in the modern world, still a mutualist.

3. J. McNicholas, A. Gilbey, A. Rennie, S. Ahmedzai, J.-A. Dono, and E. Ormerod, "Pet Ownership and Human Health: A Brief Review of Evidence and Issues," *BMJ* 331, no. 7527 (2005): 1252–1254.

4. The parasite was first discovered in Tunis, Tunisia, by researchers from the Pasteur Institute. They found the parasite in a rodent called the common

gundi (*Ctenodactylus gundi*). The gundis were being studied because they harbored *Leishmania* parasites. It was the *Leishmania* parasites that researchers were searching for when *Toxoplasma gondii* was found. *Gundi* appears to be the North African Arabic word for these rodents. The name *Toxoplasma* is from Greek, where *toxo* means "bow" and *plasma* means "shaped," in reference to the bow shape of the parasite. The name *Toxoplasma gondii*, then, full of history, means something like bow-shaped parasite from the common gundi.

5. J. Hay, P. P. Aitken, and M. A. Arnott, "The Influence of Congenital *Toxoplasma* Infection on the Spontaneous Running Activity of Mice," *Zeitschrift für Parasitenkunde* 71, no. 4 (1985): 459–462.

6. Indeed, virtually all, or perhaps even all, mammals that have so far been studied.

7. Of the phylum Apicomplexa, which also contains the malaria parasite, *Plasmodium*.

8. For the staying power of these parasites, see A. Dumètre and M. L. Dardé, "How to Detect *Toxoplasma gondii* Oocysts in Environmental Samples?" *FEMS Microbiology Reviews* 27, no. 5 (2003): 651–661.

9. And they are not alone. In work led by Amy Savage, we've found that litter boxes contain hundreds of unusual, poorly studied species.

10. In Europe, between 1 and 10 in 10,000 newborn babies are infected with *Toxoplasma gondii*. One to 2 percent will die or develop learning difficulties, and 4 to 27 percent develop retinal lesions that lead to vision impairment. See A. J. C. Cook, R. Holliman, R. E. Gilbert, W. Buffolano, J. Zufferey, E. Petersen, P. A. Jenum, W. Foulon, A. E. Semprini, and D. T. Dunn, "Sources of *Toxoplasma* Infection in Pregnant Women: European Multicentre Case-Control Study," *BMJ* 321, no. 7254 (2000): 142–147.

11. The blood of forty-one of the participants was studied in more detail using more expensive immunological assays. Those assays confirmed the results of the simpler antigen test.

12. Which is to say that being infected by a brain-manipulating parasite might make you less likely to be a department head or dean. I would have thought the opposite.

13. K. Yereli, I. C. Balcioğlu, and A. Özbilgin, "Is *Toxoplasma gondii* a Potential Risk for Traffic Accidents in Turkey?" *Forensic Science International* 163, no. 1 (2006): 34–37.

14. J. Flegr and I. Hrdý, "Evolutionary Papers: Influence of Chronic Toxoplasmosis on Some Human Personality Factors," *Folia Parasitologica* 41 (1994): 122–126.

15. J. Flegr, J. Havlícek, P. Kodym, M. Malý, and Z. Smahel, "Increased Risk of Traffic Accidents in Subjects with Latent Toxoplasmosis: A Retrospective Case-Control Study," *BMC Infectious Diseases* 2, no. 1 (2002): 11.

16. The effect of mice on stored grains was great, so great that some of our modern grains are tough because tough grains were more likely to survive being eaten by mice. See C. F. Morris, E. P. Fuerst, B. S. Beecher, D. J. Mclean, C. P. James, and H. W. Geng, "Did the House Mouse (*Mus musculus* L.) Shape the Evolutionary Trajectory of Wheat (*Triticum aestivum* L.)?" *Ecology and Evolution* 3, no. 10 (2013): 3447–3454.

17. Early agriculturalists often inadvertently offered parasites to the afterlife. See M. L. C. Gonçalves, A. Araújo, and L. F. Ferreira, "Human Intestinal Parasites in the Past: New Findings and a Review," *Memórias do Instituto Oswaldo Cruz* 98 (2003): 103–118.

18. J.-D. Vigne, J. Guilaine, K. Debue, L. Haye, and P. Gérard, "Early Taming of the Cat in Cyprus," *Science* 304, no. 5668 (2004): 259.

19. J. P. Webster, "The Effect of *Toxoplasma gondii* and Other Parasites on Activity Levels in Wild and Hybrid *Rattus norvegicus*," *Parasitology* 109, no. 5 (1994): 583–589.

20. See M. Berdoy, J. P. Webster, and D. W. Macdonald, "Parasite-Altered Behaviour: Is the Effect of *Toxoplasma gondii* on *Rattus norvegicus* Specific?" *Parasitology* 111, no. 4 (1995): 403–409.

21. E. Prandovszky, E. Gaskell, H. Martin, J. P. Dubey, J. P. Webster, and G. A. McConkey, "The Neurotropic Parasite *Toxoplasma gondii* Increases Dopamine Metabolism," *PloS One* 6, no. 9 (2011): e23866.

22. See V. J. Castillo-Morales, K. Y. Acosta Viana, E. D. S. Guzmán-Marín, M. Jiménez-Coello, J. C. Segura-Correa, A. J. Aguilar-Caballero, and A. Ortega-Pacheco, "Prevalence and Risk Factors of *Toxoplasma gondii* Infection in Domestic Cats from the Tropics of Mexico Using Serological and Molecular Tests," *Interdisciplinary Perspectives on Infectious Diseases* 2012 (2012): 529108.

23. E. F. Torrey and R. H. Yolken, "The Schizophrenia–Rheumatoid Arthritis Connection: Infectious, Immune, or Both?" *Brain, Behavior, and Immunity* 15, no. 4 (2001): 401–410.

24. J. P. Webster, P. H. L. Lamberton, C. A. Donnelly, E. F. Torrey, "Parasites as Causative Agents of Human Affective Disorders? The Impact of Anti-Psychotic, Mood-Stabilizer and Anti-Parasite Medication on *Toxoplasma gondii*'s Ability to Alter Host Behaviour," *Proceedings of the Royal Society B: Biological Sciences* 273, no. 1589 (2006): 1023–1030.

25. D. W. Niebuhr, A. M. Millikan, D. N. Cowan, R. Yolken, Y. Li, and N. S. Weber, "Selected Infectious Agents and Risk of Schizophrenia among US Military Personnel," *American Journal of Psychiatry* 165, no. 1 (2008): 99–106.

26. R. H. Yolken, F. B. Dickerson, and E. Fuller Torrey, "*Toxoplasma* and Schizophrenia," *Parasite Immunology* 31, no. 11 (2009): 706–715.

27. C. Poirotte, P. M. Kappeler, B. Ngoubangoye, S. Bourgeois, M. Moussodji, and M. J. Charpentier, "Morbid Attraction to Leopard Urine

in *Toxoplasma*-Infected Chimpanzees," *Current Biology* 26, no. 3 (2016): R98–R99.

28. Thus, infection could explain the behavior of men with too many pet cats, but not that of women with too many cats. See J. Flegr, "Influence of Latent *Toxoplasma* Infection on Human Personality, Physiology and Morphology: Pros and Cons of the *Toxoplasma*–Human Model in Studying the Manipulation Hypothesis," *Journal of Experimental Biology* 216, no. 1 (2013): 127–133.

29. Though not everywhere. In China, where until recently the keeping of cats as pets was rare, the prevalence of *Toxoplasma gondii* antibodies (and hence exposures) was also very low. It is in such countries where it may be easiest to study the consequences of infection with *Toxoplasma gondii* on particular maladies inasmuch as changes in infection status will be much easier to document. See E. F. Torrey, J. J. Bartko, Z. R. Lun, and R. H. Yolken, "Antibodies to *Toxoplasma gondii* in Patients with Schizophrenia: A Meta-Analysis," *Schizophrenia Bulletin* 33, no. 3 (2007): 729–736. doi:10.1093/schbul/sbl050.

30. M. S. Thoemmes, D. J. Fergus, J. Urban, M. Trautwein, and R. R. Dunn, "Ubiquity and Diversity of Human-Associated Demodex Mites," *PLoS One* 9, no. 8 (2014): e106265.

31. And, well, also, it wasn't the only thing Meredith was working on during those years.

32. See, for example, F. J. Márquez, J. Millán, J. J. Rodriguez-Liebana, I. Garcia-Egea, and M. A. Muniain, "Detection and Identification of *Bartonella* sp. in Fleas from Carnivorous Mammals in Andalusia, Spain," *Medical and Veterinary Entomology* 23, no. 4 (2009): 393–398.

33. A. C. Y. Lee, S. P. Montgomery, J. H. Theis, B. L. Blagburn, and M. L. Eberhard, "Public Health Issues Concerning the Widespread Distribution of Canine Heartworm Disease," *Trends in Parasitology* 26, no. 4 (2010): 168–173.

34. R. S. Desowitz, R. Rudoy, and J. W. Barnwell, "Antibodies to Canine Helminth Parasites in Asthmatic and Nonasthmatic Children," *International Archives of Allergy and Immunology* 65, no. 4 (1981): 361–366.

35. Nor is this effect of dogs on the species that live with us in our homes new. Jean-Bernard Huchet, the entomologist charged with ensuring the well-being of the mummies at the Musée de l'Homme in Paris, the guardian of their afterlives, recently dissected a dog mummy from the Egyptian site of El Deir (not far from Cairo in the Nile delta, 332 to 30 BCE). One of the dogs had date pits and figs in its stomach, indications that the dog depended on the fruit of the human settlement. The ears of the dog were also covered in brown dog ticks (*Rhipicephalus sanguineus*), a species that has now spread globally with the spread of dogs. The ticks were very likely to carry inside their bodies pathogens that could be vectored to humans; nearly a dozen different pathogens have

been found in this species of tick. All of these species were to some extent brought into the cities and homes of Egyptians via the dogs. See J. B. Huchet, C. Callou, R. Lichtenberg, and F. Dunand, "The Dog Mummy, the Ticks and the Louse Fly: Archaeological Report of Severe Ectoparasitosis in Ancient Egypt," *International Journal of Paleopathology* 3, no. 3 (2013): 165–175.

36. Among them species of *Arthrobacter*, *Sphingomonas*, and *Agrobacterium*.

37. A. A. Madden, A. Barberán, M. A. Bertone, H. L. Menninger, R. R. Dunn, and N. Fierer, "The Diversity of Arthropods in Homes across the United States as Determined by Environmental DNA Analyses," *Molecular Ecology* 25, no. 24 (2016): 6214–6224; M. Leong, M. A. Bertone, A. M. Savage, K. M. Bayless, R. R. Dunn, and M. D. Trautwein, "The Habitats Humans Provide: Factors Affecting the Diversity and Composition of Arthropods in Houses," *Scientific Reports* 7, no. 1 (2017): 15347.

38. C. Pelucchi, C. Galeone, J. F. Bach, C. La Vecchia, and L. Chatenoud, "Pet Exposure and Risk of Atopic Dermatitis at the Pediatric Age: A Meta-Analysis of Birth Cohort Studies," *Journal of Allergy and Clinical Immunology* 132 (2013): 616–622.e7.

39. K. C. Lødrup Carlsen, S. Roll, K. H. Carlsen, P. Mowinckel, A. H. Wijga, B. Brunekreef, M. Torrent, et al., "Does Pet Ownership in Infancy Lead to Asthma or Allergy at School Age? Pooled Analysis of Individual Participant Data from 11 European Birth Cohorts," *PLoS One* 7 (2012): e43214.

40. G. Wegienka, S. Havstad, H. Kim, E. Zoratti, D. Ownby, K. J. Woodcroft, and C. C. Johnson, "Subgroup Differences in the Associations between Dog Exposure During the First Year of Life and Early Life Allergic Outcomes," *Clinical and Experimental Allergy* 47, no. 1 (2017): 97–105.

41. S. J. Song, C. Lauber, E. K. Costello, C. A. Lozupone, G. Humphrey, D. Berg-Lyons, J. G. Caporaso, et al., "Cohabiting Family Members Share Microbiota with One Another and with Their Dogs," *Elife* 2 (2013): e00458; M. Nermes, K. Niinivirta, L. Nylund, K. Laitinen, J. Matomäki, S. Salminen, and E. Isolauri, "Perinatal Pet Exposure, Faecal Microbiota, and Wheezy Bronchitis: Is There a Connection?" *ISRN Allergy* 2013 (2013).

42. M. G. Dominguez-Bello, E. K. Costello, M. Contreras, M. Magris, G. Hidalgo, N. Fierer, and R. Knight, "Delivery Mode Shapes the Acquisition and Structure of the Initial Microbiota across Multiple Body Habitats in Newborns," *Proceedings of the National Academy of Sciences* 107, no. 26 (2010): 11971–11975.

## CHAPTER 11

1. Also called 52 or 52a.

2. Or at least any other microbe in those countries in which public health systems, waste treatment, and hand washing were well established. H. R.

Shinefield, J. C. Ribble, M. Boris, and H. F. Eichenwald, "Bacterial Interference: Its Effect on Nursery-Acquired Infection with *Staphylococcus aureus*. I. Preliminary Observations on Artificial Colonization of Newborns," *American Journal of Diseases of Children* 105 (1963): 646–654.

3. Based on the most recent estimates, just a few decades earlier. See P. R. McAdam, K. E. Templeton, G. F. Edwards, M. T. G. Holden, E. J. Feil, D. M. Aanensen, H. J. A. Bargawi, et al., "Molecular Tracing of the Emergence, Adaptation, and Transmission of Hospital-Associated Methicillin-Resistant *Staphylococcus aureus*," *Proceedings of the National Academy of Sciences* 109, no. 23 (2012): 9107–9112.

4. They had previously suggested that the answer, in the case of such infections, was the detailed study of the biology of the pathogen. Now they were being called to task. See H. F. Eichenwald and H. R. Shinefield, "The Problem of Staphylococcal Infection in Newborn Infants," *Journal of Pediatrics* 56, no. 5 (1960): 665–674.

5. Shinefield et al., "Bacterial Interference: Its Effect On Nursery-Acquired Infection," 646–654.

6. H. R. Shinefield, J. C. Ribble, M. B. Eichenwald, and J. M. Sutherland, "V. An Analysis and Interpretation," *American Journal of Diseases of Children* 105, no. 6 (1963): 683–688.

7. These were the very same bacteria my colleagues and I would later find dominated belly buttons. See J. Hulcr, A. M. Latimer, J. B. Henley, N. R. Rountree, N. Fierer, A. Lucky, M. D. Lowman, and R. R. Dunn, "A Jungle in There: Bacteria in Belly Buttons Are Highly Diverse, but Predictable," *PLoS One* 7, no. 11 (2012): e47712.

8. It was also possible that other species of bacteria, such as species of *Micrococcus* or *Corynebacterium*, might help repel 80/81, but Eichenwald and Shinefield thought that the competition between related species would be more intense than that between less-related species. In this, skin microbes are like plant species in grasslands or forests. More closely related plants tend to be more ecologically similar and more likely to compete with and be able to exclude each other. See J. H. Burns and S. Y. Strauss, "More Closely Related Species Are More Ecologically Similar in an Experimental Test," *Proceedings of the National Academy of Sciences* 108, no. 13 (2011): 5302–5307.

9. D. Janek, A. Zipperer, A. Kulik, B. Krismer, and A. Peschel, "High Frequency and Diversity of Antimicrobial Activities Produced by Nasal *Staphylococcus* Strains against Bacterial Competitors," *PLoS Pathogens* 12, no. 8 (2016): e1005812.

10. Among ants, for example, a classic example of interference competition is when ants of the species *Novomessor cockerelli* interfere with the foraging of their competitors, species of *Pogonomyrmex harvester* ants, by plugging the nest entrances of the latter with stones.

11. One exception being René Dubos. H. L. Van Epps, "René Dubos: Unearthing Antibiotics," *Journal of Experimental Medicine* 203, no. 2 (2006): 259.

12. Shinefield et al., "Bacterial Interference: Its Effect on Nursery-Acquired Infection," 646–654.

13. This is work done by the extraordinary scientist with the superhero name, Paul Planet, and his collaborators. D. Parker, A. Narechania, R. Sebra, G. Deikus, S. LaRussa, C. Ryan, H. Smith, et al., "Genome Sequence of Bacterial Interference Strain *Staphylococcus aureus* 502A," *Genome Announcements* 2, no. 2 (2014): e00284-14.

14. This same concept, that success is best predicted by the number of individuals introduced (or the number of attempts at introduction), also holds for other sorts of colonization. For example, one of the best predictors of whether an introduced ant species succeeds in establishing is how many times it was introduced. See A. V. Suarez, D. A. Holway, and P. S. Ward, "The Role of Opportunity in the Unintentional Introduction of Nonnative Ants," *Proceedings of the National Academy of Sciences of the United States of America* 102, no. 47 (2005): 17032–17035.

15. Interestingly, in those few cases when 502A didn't take, it was often because other *Staphylococcus* had already colonized the noses and belly buttons of the babies. See Shinefield et al., "Bacterial Interference: Its Effect on Nursery-Acquired Infection," 646–654.

16. H. R. Shinefield, J. M. Sutherland, J. C. Ribble, and H. F. Eichenwald, "II. The Ohio Epidemic," *American Journal of Diseases of Children* 105, no. 6 (1963): 655–662.

17. H. R. Shinefield, M. Boris, J. C. Ribble, E. F. Cale, and Heinz F. Eichenwald, "III. The Georgia Epidemic," *American Journal of Diseases of Children* 105, no. 6 (1963): 663–673. Also see M. Boris, H. R. Shinefield, J. C. Ribble, H. F. Eichenwald, G. H. Hauser, and C. T. Caraway, "IV. The Louisiana Epidemic," *American Journal of Diseases of Children* 105, no. 6 (1963): 674–682.

18. H. F. Eichenwald, H. R. Shinefield, M. Boris, and J. C. Ribble, "'Bacterial Interference' and Staphylococcic Colonization in Infants and Adults," *Annals of the New York Academy of Sciences* 128, no. 1 (1965): 365–380.

19. D. Janek, A. Zipperer, A. Kulik, B. Krismer, and A. Peschel, "High Frequency and Diversity of Antimicrobial Activities Produced by Nasal *Staphylococcus* Strains against Bacterial Competitors," *PLoS Pathogens* 12, no. 8 (2016): e1005812.

20. This is what Paul Planet thinks may be going on.

21. C. S. Elton, *The Ecology of Invasions by Animals and Plants* (London: Methuen & Co, 1958).

22. For the quote, see J. D. van Elsas, M. Chiurazzi, C. A. Mallon, D. Elhottovā, V. Krištůfek, and J. F. Salles, "Microbial Diversity Determines the

Invasion of Soil by a Bacterial Pathogen," *Proceedings of the National Academy of Sciences* 109, no. 4 (2012): 1159–1164. For a general review, see J. M. Levine, P. M. Adler, and S. G. Yelenik, "A Meta-Analysis of Biotic Resistance to Exotic Plant Invasions," *Ecology Letters* 7, no. 10 (2004): 975–989.

23. J. M. H. Knops, D. Tilman, N. M. Haddad, S. Naeem, C. E. Mitchell, J. Haarstad, M. E. Ritchie, et al., "Effects of Plant Species Richness on Invasion Dynamics, Disease Outbreaks, and Insect Abundances and Diversity," *Ecology Letters* 2 (1999): 286–293.

24 J. D. van Elsas, M. Chiurazzi, C. A. Mallon, D. Elhottovā, V. Krištůfek, and J. F. Salles, "Microbial Diversity Determines the Invasion of Soil by a Bacterial Pathogen," *Proceedings of the National Academy of Sciences* 109, no. 4 (2012): 1159–1164.

25. Nor were the results of van Elsas and his colleagues a fluke of choosing to work on *E. coli*. Results have been similar in studies considering the invasion of the soil around wheat roots by the bacterial species *Pseudomonas aeruginosa*. See A. Matos, L. Kerkhof, and J. L. Garland, "Effects of Microbial Community Diversity on the Survival of *Pseudomonas aeruginosa* in the Wheat Rhizosphere," *Microbial Ecology* 49 (2005): 257–264.

26. We often look back on choices societies have made and wonder whether anyone sounded the alarm when a bad decision was being made. We are quick to suggest that decades, or centuries, or millennia ago, our antecedents didn't know enough to choose more wisely. In this particular case, we did know enough. In 1965, Shinefield and Eichenwald clearly laid out the problems that would emerge were we to focus completely on antibiotics. See Shinefield et al., "V. An Analysis and Interpretation," 683–688.

27. Fleming said, "There is the danger that the ignorant man may easily underdose himself and by exposing his microbes to non-lethal quantities of the drug make them resistant. Here is a hypothetical illustration. Mr. X. has a sore throat. He buys some penicillin and gives himself, not enough to kill the streptococci but enough to educate them to resist penicillin. He then infects his wife. Mrs. X gets pneumonia and is treated with penicillin. As the streptococci are now resistant to penicillin the treatment fails. Mrs. X dies. Who is primarily responsible for Mrs. X's death? Why Mr. X whose negligent use of penicillin changed the nature of the microbe."

28. M. Baym, T. D. Lieberman, E. D. Kelsic, R. Chait, R. Gross, I. Yelin, and R. Kishony, "Spatiotemporal Microbial Evolution on Antibiotic Landscapes," *Science* 353, no. 6304 (2016): 1147–1151.

29. F. D. Lowy, "Antimicrobial Resistance: The Example of *Staphylococcus aureus*," *Journal of Clinical Investigation* 111, no. 9 (2003): 1265.

30. E. Klein, D. L. Smith, and R. Laxminarayan, "Hospitalizations and Deaths Caused by Methicillin-Resistant *Staphylococcus aureus*, United States, 1999–2005," *Emerging Infectious Diseases* 13, no. 12 (2007): 1840.

31. Just why the use of antibiotics leads cows and pigs to grow more quickly is not entirely understood.

32. S. S. Huang, E. Septimus, K. Kleinman, J. Moody, J. Hickok, T. R. Avery, J. Lankiewicz, et al., "Targeted versus Universal Decolonization to Prevent ICU Infection," *New England Journal of Medicine* 368, no. 24 (2013): 2255–2265.

33. R. Laxminarayan, P. Matsoso, S. Pant, C. Brower, J.-A. Røttingen, K. Klugman, and S. Davies, "Access to Effective Antimicrobials: A Worldwide Challenge," *Lancet* 387, no. 10014 (2016): 168–175. For more on policy solutions to the resistance challenge, see P. S. Jorgensen, D. Wernli, S. P. Carroll, R. R. Dunn, S. Harbarth, S. A. Levin, A. D. So, M. Schluter, and R. Laxminarayan, "Use Antimicrobials Wisely," *Nature* 537, no. 7619 (2016); K. Lewis, "Platforms for Antibiotic Discovery," *Nature Reviews Drug Discovery* 12 (2013): 371–387.

## CHAPTER 12

1. D. E. Beasley, A. M. Koltz, J. E. Lambert, N. Fierer, and R. R. Dunn, "The Evolution of Stomach Acidity and Its Relevance to the Human Microbiome," *PloS One* 10, no. 7 (2015): e0134116.

2. G. Campbell-Platt, *Fermented Foods of the World. A Dictionary and Guide* (Oxford: Butterworth Heinemann, 1987).

3. Kimchi is much more diverse than are most other fermented foods. Not only can individual kimchi types contain tens or even hundreds of species (that differ from one person's kimchi to another person's kimchi, it seems) but also different types of kimchi have very different microbes. See E. J. Park, J. Chun, C. J. Cha, W. S. Park, C. O. Jeon, and J. W. Jin-Woo Bae, "Bacterial Community Analysis During Fermentation of Ten Representative Kinds of Kimchi with Barcoded Pyrosequencing," *Food Microbiology* 30, no. 1 (2012): 197–204. In addition to *Staphylococcus* and *Lactobacillus*, common genera of bacteria in kimchi include *Leuconostoc* and a close relative of *Leuconostoc*, *Weisella* (both of which we find to be abundant in refrigerators), *Enterobacter* (a fecal microbe), and *Pseudomonas*.

4. *Bacillus subtilis*, the same bacterial species that makes feet stinky (and that was found in abundance on the International Space Station). For more on Korean fermentation, see J. K. Patra, G. Das, S. Paramithiotis, and H.S. Shin, "Kimchi and Other Widely Consumed Traditional Fermented Foods of Korea: A Review," *Frontiers in Microbiology* 7 (2016).

5. I highly recommend viewing the 1903 documentary film *Cheese Mites* (produced by Charles Urban and directed by F. Martin Duncan), a film that

highlights the beauty of an animal that can help turn one food into another. www.youtube.com/watch?v=wR2DystgByQ.

6. L. Manunza, "Casu Marzu: A Gastronomic Genealogy," in *Edible Insects in Sustainable Food Systems* (Cham, Switzerland: Springer International, 2018).

7. For a nice description of the early history of bread and a tale of the quest to re-create ancient bread technology, see E. Wood, *World Sourdoughs from Antiquity* (Berkeley, CA: Ten Speed Press, 1996).

8. These breads were a kind of money, a ration, and, like beer, a unit of exchange. Bread baking was a way of turning a hard-to-work grain into an easy-to-store, -trade, -sell, or -eat food. See D. Samuel, "Bread Making and Social Interactions at the Amarna Workmen's Village, Egypt," *World Archaeology* 31, no. 1 (1999): 121–144.

9. Nor has the question even been very seriously studied. No one, for instance, has searched for ancient DNA in any of the many mummified breads that accompany Egyptian burials. Such burials have already told us so very much about daily life in antiquity. They have much more to tell, though I'm not sure this is the afterlife the Egyptians had in mind.

10. The details of this process vary. Some use only distilled water, others only rainwater. Bakers also vary in terms of the kind of flour they use, the temperature at which their starters are kept, and even whether other microbe-laden foods (including fruits) are added to the mix.

11. L. De Vuyst, H. Harth, S. Van Kerrebroeck, and F. Leroy, "Yeast Diversity of Sourdoughs and Associated Metabolic Properties and Functionalities," *International Journal of Food Microbiology* 239 (2016): 26–34.

12. One study of bakeries found that even though the flour being used contained bacteria of the genus *Enterobacter* (a potentially pathogenic fecal microbe), these bacteria never established in the sourdough starter. They were killed, it seems, by the sourdough bacteria and the acid they had produced. In the same study, the bacteria were very diverse in the flour, the mixing bowl, and even the bread storage box—but not in the sourdough, where a simple, stable microbial garden grew.

13. When refrigerators and freezers were invented, they were a new, alternative way of storing food, but mostly they are less effective than is fermentation. When you buy food, all of your food is filled with microbes (even the vacuum-sealed food). When you put your food in the refrigerator, it slows down the feeding and reproduction of the microbes in the food. The "best by" label on the food in your fridge is essentially a measure of how long it takes those microbes in your food, even under cold conditions, to divide and metabolize enough to take over the food. This is what the "best by" label should really say: "Not totally thick with microbes until January 4," though just how

long you get actually depends on which microbes have colonized your food from your kitchen, hands, and breath each time you open the jar. In other words, "best by January 4" is a lie, but a good rule-of-thumb lie, one that helps us get through the day alive.

14. Sometimes these breads are made sour by adding a strain of *Lactobacillus reuteri* originally from rodent feces. If you don't believe me, read M. S. W. Su, P. L. Oh, J. Walter, and M. G. Gänzle, "Intestinal Origin of Sourdough *Lactobacillus reuteri* Isolates as Revealed by Phylogenetic, Genetic, and Physiological Analysis," *Applied and Environmental Microbiology* 78, no. 18 (2012): 6777–6780.

15. In doing so, *Saccharomyces cerevisiae* seems to only rarely be part of the starter community, though our picture is biased. It appears that once packaged yeast is used in a bakery, it easily becomes part of the indoor yeast community in the bakery (setting up shop on the mixers, in the flour, in the storage containers, and so on) and, in doing so, readily "contaminates" new starters. This doesn't prevent the starters from doing their job, but it does reduce the diversity of starters, a more subtle element of the microbial homogenization set in motion by the industrial-scale production and use of *Saccharomyces*. See F. Minervini, A. Lattanzi, M. De Angelis, G. Celano, and M. Gobbetti, "House Microbiotas as Sources of Lactic Acid Bacteria and Yeasts in Traditional Italian Sourdoughs," *Food Microbiology* 52 (2015): 66–76.

16. No guess as to what might have made Herman turn pink. It probably had nothing to do with the earthquake.

17. We wanted to avoid letting them feed the starters before we sampled because if they fed the starters in the kitchen (and they would), they would likely inadvertently introduce microbes from the kitchen into the starters. This would happen. It was unavoidable, but by sampling before it happened, we had our best chance of measuring the microbes unique to each baker's efforts, body, and home.

18. We controlled for the big differences, but it took constant work. We even had to keep an eye out to make sure the bakers didn't introduce other ingredients into the breads, ingredients they were desperate to add (and that seemed to appear miraculously from pockets and smocks): "But what about a little garlic? Just a little? How about some sesame!"

19. D. A. Jensen, D. R. Macinga, D. J. Shumaker, R. Bellino, J. W. Arbogast, and D. W. Schaffner, "Quantifying the Effects of Water Temperature, Soap Volume, Lather Time, and Antimicrobial Soap as Variables in the Removal of *Escherichia coli* ATCC 11229 from Hands," *Journal of Food Protection* 80, no. 6 (2017): 1022–1031.

20. A. A. Ross, K. Muller, J. S. Weese, and J. Neufeld, "Comprehensive Skin Microbiome Analysis Reveals the Uniqueness of Human-Associated Microbial Communities among the Class Mammalia," *bioRxiv* (2017): 201434.

21. N. Fierer, M. Hamady, C. L. Lauber, and R. Knight, "The Influence of Sex, Handedness, and Washing on the Diversity of Hand Surface Bacteria," *Proceedings of the National Academy of Sciences* 105, no. 46 (2008): 17994–17999.

22. A. Döğen, E. Kaplan, Z. Öksüz, M. S. Serin, M. Ilkit, and G. S. de Hoog, "Dishwashers Are a Major Source of Human Opportunistic Yeast-Like Fungi in Indoor Environments in Mersin, Turkey," *Medical Mycology* 51, no. 5 (2013): 493–498.

# INDEX

**Rob Dunn** is a professor in the Department of Applied Ecology at North Carolina State University and in the Natural History Museum of Denmark at the University of Copenhagen. He is also the author of five books. He lives in Raleigh, North Carolina.

*Credit: Amanda Ward*